LGM

DEV GUALTIERI

Tikalon Press
http://www.tikalonpress.com

ISBN: 978-1-942459-02-6

*Cover background image from **KStars***
(https://edu.kde.org/kstars/)

*Cover Font, **nasalization**, by Typodermic Fonts Inc. (typodermicfonts.com)*

AUTHOR'S FORWARD

LGM is a novel set in the present. The year is not explicitly stated, but the time period could be any time at the start of the twenty-first century.

The detection of an intelligent extraterrestrial signal sidetracks the routine existence of a radio astronomer, a reclusive mathematician and two computer entrepreneurs.

The discovery and analysis of this intelligent signal from the extraterrestrials, affectionately called the "Little Green Men," proceeds as it might in real life.

This is a work of fiction. All characters appearing in this work are fictitious. Any resemblance to real persons, living or dead, is purely coincidental.

ACKNOWLEDGEMENTS

Special thanks to my son, daughter, and son-in-law for their careful reading of the manuscript. They discovered quite a few typographical errors, and they also suggested improvements to the structure and content of this book.

CONTENTS

PROLOGUE - GREECE, 450 BC

It was a fine morning, the beginning of a fine day, so I was not surprised when a messenger summoned me to dine, mid-day, with Leucippus. I immediately assented, since the Old Boy always had something to capture my attention. Most men live in the world without thinking at all about it, living from hand to mouth without questioning why things are as they are, and not otherwise. Leucippus was not one of them.

Leucippus had lectured me and our small circle about how different things might arise from more fundamental substances. While one can hack away at a thing, cutting it into smaller and smaller pieces, might there not be a limit? There may be something that describes the essence of a thing, something that cannot be cut without a loss of that essence. This idea of the a-tom, the thing that cannot be cut, was his most interesting idea, yet. Whatever was on his mind, now, might be more interesting, still.

I passed the morning in the usual manner, tending to the garden, and leaving a small devotion to Athena, who had guided me well in her arts. She inspired me in the design of the garden cistern that held water between the rains and allowed a greater harvest from that small plot than I had ever had.

My shortened shadow marked the time for my walk to the house of Leucippus, a mere ten stadia from my own. Although the Sun was at mid-day, the heat was not at all oppressive, and it was a welcome change from the typical chill of the season. Leucippus and two other friends were there to greet me upon my arrival, all sitting on stools near the walkway. I was immediately offered water, and Leucippus, my host, was the first to speak.

"Peleus, old friend, have the gods been good?"

"Yes, they have, but it's strange to hear your talking about the gods. We were led to believe that the gods are not a part of your world."

"You misunderstand. I'm the first to admit that there are forces, above and about, that control our destiny. It's how we name them that makes us different. But, it's much better to eat and talk, than to just talk."

At that, he signaled the servants to bring us food. When my beverage was at hand, I poured a libation to the gods, as did all those seated. Leucippus offered a libation also, perhaps from habit. As senior in the company, next to our host, it fell to me to begin the conversation.

"Leucippus, will you once again regale us with more details of the structure of the world about us, as defined in your philosophy; or, have other things been on your mind?"

"Peleus how nice it is for you to ask, for what would I be if not for my ideas. I am certainly no good in the vineyard, and I have no crafts to match your own. What concerns me now is neither the composition of matter, nor the nature of the gods, but the nature of people."

Theodoros, another from the group, spoke.

"Do you mean the customs of men, which differ widely as we travel from place to place?"

Leucippus smiled broadly in that way he had to indicate he meant something far more fundamental.

"As you say, we travel, and we find strange people in the distant lands. We hear also the tales of those who have traveled farther still, and the people they have met. Do you wonder how there are always people wherever we go?"

I made answer for the group.

"Why would you not expect there to be people everywhere? If a man lives at one place, he can just as well pack his household and move a few hundred stadia farther still. I see no problem with that."

Leucippus smiled broadly, again, as if I had fallen into his trap.

"If he moves, say, to a place where there grows less barley, but more spelt, he would survive, would he not? Or, if he finds a place high atop a mountain, would he not just wear furs to keep himself warm?"

"Yes, that is all logical. Men will adapt to their environment in order to survive, just as an old man will use a stick to help him to walk."

"And what of the other worlds?"

"Other worlds?"

"Earth is just one of many worlds. It has its Sun to provide heat and light. The stars in the sky are other suns providing heat and light to their worlds."

I felt a need to object.

"There is no evidence for these other worlds! Things on Earth we can see, but things beyond the Earth will always be unknown."

"You do not see the gods, but yet you believe. Believe with me, for the moment, that there are other worlds. Is it not logical that therein live men who are adapted to their worlds?"

This was just like Leucippus, to state the ridiculous and render it logical.

"Yes, I agree that there may be men on other worlds. However, you must admit that, since there would be no means to discover their existence, it might as well be that they are not really there."

1 NOISE

Although their offices were on opposite sides of the building, the trip from Brian's office to Tukey's office generally took just a minute. Brian, however, was extending the trip to a much longer period. He stopped to read some of the bulletin board notices, which was something he rarely did. All useful information came from personal encounters, email, and the Internet. What he may have missed beyond those was likely not important. Still, someone had taken the time to post these items, so one or another might prove to be important. This time, that wasn't the case.

Brian paused to stare through the corridor skylight and admire the huge radio telescope dish responsible for the signal traces he carried in his hands. Such a huge structure would be a marvel even to someone who didn't understand its purpose. It was a megalith of the modern age. He felt privileged to be a part of such a venture. Somehow, civilization had advanced to the stage at which humans could afford to build such "useless" devices, and even pay people like him to operate them.

Arriving at Tukey's office, Brian stood at the closed office door for a few seconds. He really hated to disturb the Old Man; but, that's why the Old Man was there, wasn't it? Brian had spent the last few minutes convincing himself that what he found demanded Tukey's attention. All he needed to do was to follow through. He knocked and opened the door in response to the grunt from the other side. Tukey was in his usual repose, leaning back in his office chair with a journal in hand.

"Professor, do you have time to look at a signal trace?"

Tukey looked up from his journal to see Brian, one of his post-docs, peering through the doorway. The journal he was reading was the traditional paper type. Tukey had never made the transition to reading research papers on a computer screen or tablet. Some of his colleagues extolled one feature of tablets, the ability to enlarge the type font to make for easier reading. Tukey didn't think he had reached the point at which that was necessary, and it would be somewhat strange reading an astrophysics paper in type as large as that in a child's storybook.

Digital publication did have another advantage – faster distribution. Although the journal he held was the current number, it contained research that must be many months, if not years, old. It didn't matter. In his field, if anything important did happen, he would hear about it from a television newscast, not from a journal.

"Sure, let's see what you've got."

Brian, a recently minted PhD from the University of Wisconsin, Madison, spread a meter-long chart trace over several stacks of books and papers on the table behind Tukey's desk. Old scientists always had a lot of paper in their offices. Brian's office was nearly bare. He had used the large format printer to allow a closer examination of the raw data, and he had carefully trimmed the excess borders to make the graph easier to handle and read. It didn't occur to young Brian that the data looked like a traditional strip-chart recorder pen trace from Tukey's own student days.

Tukey immediately saw why the chart was interesting, but he waited for Brian to point it out. The aged professor was acting paternally. Scientist and apprentice, father and son, this type of relationship always seemed to work, both to the benefit of the individuals, and society as a whole. Society had signed onto this particular scientist-apprentice team wholeheartedly, funding Tukey, his post-doc, Brian, and their associated overhead, at a million dollars a year.

"This is one channel from the channelizer. It doesn't matter which one - they all look about the same. If you compare this segment with what comes before and after, you see that the character of the noise changes slightly in sort of a bell curve. The power level doesn't seem to change, but if you do an FFT of the raw data, there's a definite excess of high frequency noise."

There was a qualitative difference, just barely visible to the eye, but Brian must have been alerted to it by the automated signal analysis programs. The programs must have been well crafted to pluck this little patch of scruff from a sea of debris. Tukey looked sideways along the paper as a way to accentuate the profile of the signal trace, looking for any sort of bump. There wasn't any, but the signal was different in the way that Brian described.

"You checked the Omni."

Tukey was referring to a second receiver designed to record local radio interference.

"None of this is in the Omni data. I do see some of that motorcycle that passes a few miles to the North every night about this time, but that noise isn't mirrored in the data from Main. And, most interestingly, the width of the blip is what you would expect for a point source convoluted with the beam width."

Brian was thorough, as usual, and that's why Tukey had recruited him.

"Well, it does look different, but can we afford an immediate recheck? How's the Crummy survey going?

Hearing a professor refer to the Kumhansl Milliarcsecond Survey in the same slang as the students and staff prompted a quick smile from Brian.

"We're at least a month ahead on the data collection, and three weeks ahead on the offline analysis. The monthly report won't be a problem, barring any equipment troubles."

"So, we eat for another day. Do a recheck, this time at a slower scan rate, and open up the bandwidth."

Brian nodded assent. On his walk to Tukey's office, he had predicted the exact outcome of the discussion. That's one nice thing about science. It's so objective that consensus is rarely a problem. There is typically only one path to Truth.

"... And charge everything to Crummy."

"Crummy? How can I justify that on my time report?"

"That reminds me, I'm about six weeks behind on my own time reports."

"What?"

"Just kidding. Do you think the Bean Counters would let me get even one week behind? They like to close their books, neat and complete, at the end of every week, and then send out invoices that are ignored for ninety days. All in all, I don't think scientists should spend any time at all on this finance stuff; but, since we're forced to do it, we need to work the system."

Tukey paused, as if for emphasis.

"Brian, what makes your data look any different from some instrumental noise?"

"Well, at this point it could be anything, even a bad amplifier on the front-end. But I'm pretty sure that's not the case."

"Well, just-to-be-sure-wink-wink, we'll need to do some more measurements. After all, we want the data from the Krumhansl Survey to be beyond reproach... So, charge this to Crummy."

Brian smiled. He was learning much more on this apprenticeship than any textbook could teach. He left the office and headed towards the control room. Tukey retrieved the journal, but his gaze was still on the data trace that Brian had courteously left behind.

"Are we grasping at straws?" he thought. "Then again, that's what we're paid to do."

He read for another few minutes, and then dozed off.

Tukey rarely dreamed. There was the occasional nightmare, typically caused by too heavy a restaurant meal, but not much more. There must have been something about Brian's data trace that triggered a rare, recurrent dream. Sitting late at night, alone in the telescope control room, he detected a genuine interstellar transmission. The senders had been kind enough to send the message in English. As always, just as he started to read the message, he awoke. Maybe he should call that dream a nightmare, too.

2 VENTURE CAPITAL

Rick stared blankly out the limousine window. He should have been feeling dread, or exhilaration, or a mixture of both. Instead, he was just numb. It didn't help that traffic seemed to be at a standstill. Everyone seemed to have the need to be somewhere else at all times, and they all tried to get there in automobiles. Modern life seemed to balance every advance in the human condition with some step backwards.

"How did it come down to this, Tom?"

Tom, who was to be their point man at the meeting, was reviewing his notes for possibly the hundredth time. Tom left nothing to chance, so he made a very good business partner. Rick hoped that his own creativity added enough to the package to make it a balanced partnership.

"What do you mean?"

"You know. Here we are, on our way to stake our future on a handful of presentation slides. There's not a single equation in the bunch, and our idea is so buried in financial gobbledygook that even I can't find it. Then there's the audience, a bunch of white-haired suits who couldn't take the square root of four. The only reason they want to see us is that they have twenty million dollars in loose change and they don't know how to spend it."

"We have a good business plan. We paid the consultant to help us write a winner. He knows what they want, what they really look for. Hey, techno-geeks are only good for entertainment value in Suit Culture. The Suits value money, not science."

The limo was Tom's idea. To woo Big Money, you needed to look the part. Tom went back to his notes, and Rick continued to stare out the window. He thought back to their college days when getting to class was an adventure. They fled the dorms as soon as they could, right after freshman year, and got an apartment a few miles from campus. The distance meant they needed a car, and on their budget they needed a very inexpensive car. In the winter they had to pull the battery out every evening and keep it warm in the apartment overnight, or the car wouldn't start the next morning. Now, here they were, eight years later, in a limo.

Tom and Rick had been roommates from the first day of college. Rick majored in Physics, and Tom was in Economics. Despite their different interests, they were matched well in intelligence and were able to feed off the fringes of each other's field. Like most students, they both had mastered the family computer at an early age, and they each were able to write simple programs while still in elementary school. But they really weren't that interested in computers, or they would have studied Computer

Science. Instead, they used their mastery of computers to advance themselves in their respective fields.

Rick had an early interest in particle physics and spent his graduate school days attached to a very large team in charge of a very large detector doing a very large experiment. The data stream was tremendous, and it was his job to reduce it to human proportions, at the same time trying not to lose anything important. The experiment was exciting, but the work was almost routine. Sure, there was a lot of his intellectual input in the pattern matching algorithms (Rick couldn't have been assured of an eventual PhD without that), but he quickly tired of being just a small part of a huge team.

When there was a publication, the team members were not even individually listed as authors. Decades prior, the major Physics journals started to object to papers in which the authors' names took more page area than the actual article. They were the "Blue Ball Consortium," a not so inside joke based on the appearance of their detector.

Rick was imagining a monotonous, but comfortable life after graduation when he met his first "Rocket Scientist." "Rocket Scientists," or "Quants," are a special breed of Physicist who try to model financial markets with the same tools they use to model nature. Better knowledge of the market wasn't the only goal. That knowledge would be used to make money; and, often, a lot of money, very fast.

Rick saw a posting of a seminar with the strange title, "Random Matrix Theory as Applied to Sampling in the Analysis of Financial Derivatives." At first he thought it was a joke. By chance, the seminar was scheduled for April first, and Physicists, as Feynman proved, are very creative jokers. Maybe the professors wanted to see which of their students were dim-witted enough to show up. Well, Rick was one of the dim-witted, and he was amazed. Here was a PhD Physicist doing the same sorts of math and computer analysis Rick was doing, but he was happy in his job and making a ton of money. Sure, after the talk, some of the audience thought the guy had sold out, but Rick was hooked. After the typical span of time, Rick was granted his degree, and he was free to pursue this unusual career path.

Rick was ready, but the world was not. He wasn't the only Physicist who had discovered "Rocket Science," and the ground floor of the field was fully filled. Many of the other scientists had pursued an MBA while taking their PhD, looking for any edge to advance into the field. Many of them came from backgrounds in statistical mechanics and thermodynamics that were more relevant disciplines to financial analysis than particle physics. Rick was depressed, and he had resigned himself to a future of staring at spark chamber photographs, but a trip to the dentist changed all that.

It's not that the dentist himself, a wise old professional, had offered just the right advice, like a bartender, while performing his usual services. The inspiration came from a throw-away magazine he flipped through in the waiting room. There, in the back pages, was an ad that reminded him of the ads he saw in comic books as a child. There was a line drawing of a smiling man with a fistful of dollars radiating dollar signs from his head. In bold type were the words, "BE YOUR OWN BOSS." The ad had nothing to do with any profession that would have interested him, but it got him thinking. Soon thereafter, Tom, who was plugging away at his advanced degree at another school, and Rick hatched their plan to join forces and take finance by storm. Their first hurdle was convincing the wives.

Graduate student wives, at least the wives of technologists, are a complex breed. On the one hand, they are accustomed to a poverty lifestyle, as all student wives must be; on the other hand, they are gamblers. In their minds, this early phase of marriage is like a lottery - they're betting their husbands will get rich. Not only that, but they will get rich rather quickly.

There is, of course, the other species, the wives of non-technology graduate students. These are either masochists, or they enjoy a Bohemian lifestyle. Their husbands would most likely avoid slinging burgers, and they might even aspire to some middle-management position (probably in an amorphous field like human resources), but they've resigned themselves to the fact that there would be no Mercedes SUV in the driveway. It must be true love.

Tom's wife and Rick's wife were fully supportive and rather enjoyed the idea that they would live within a short automobile ride from each other. They would each have a shoulder to cry on, and the dinner parties would be nice.

So, there they were, Tom and Rick, in a limo, building up enough courage to ask for just a few million dollars - small bills, please. Almost as an omen, the clouds had parted ahead, letting a few rays of light into the city canyons. At the same time, traffic began to move. Rick, ever analytical, started to wonder whether there was a causal relationship at play. Perhaps the rays of the Sun had subconsciously triggered the slightest bit of optimism in the drivers, and that was enough to get the line of cars moving.

Such things were common in nature, where a small grain of sand would cause an avalanche, or a bit of dust landing in a cool pond will cause it to freeze, instantly. Just a single driver's stopping properly at a traffic signal and not rushing ahead to block cross traffic might just have this same affect. People pride themselves on their individuality, but when they're in large groups, their fate is generally governed by statistical laws they don't understand.

Rick and Tom's business plan was based on essentially that idea, that large quantities of things, when examined in the aggregate, will perform predictably. They were going to game the stock market using computer algorithms.

3 CHICKENS, CHICKENS, EVERYWHERE

Roos stared out at the expansive plain before him.

"Chickens, chickens, everywhere, nor any egg to eat..."

He remembered some of the real verses to the poem.

Water, water, everywhere,
And all the boards did shrink;
Water, water, everywhere,
Nor any drop to drink.

The notion of shrinking boards set off an uncontrollable flurry of differential equations and tensors in his head, as he subconsciously starting building a model of how boards really do shrink. He shuddered at this unwelcome onslaught and tried to block the thoughts from his consciousness. He succeeded only partially. Mathematics always operated as a background process in his brain.

"You can take the man out of mathematics, but you can never take mathematics out of the man."

Martin Roos had been a Wunderkind of mathematics. His parents had noticed his facility with numbers before he attended preschool. Since he enjoyed arranging things in groups, his parents were worried at first that he might have Asperger syndrome. From the magazine articles that they had read, what he was doing was a sure sign. Their doctor advised them, however, that all other symptoms were absent. Perhaps their son was a budding artist who enjoyed changing the juxtaposition of colored objects. They shouldn't even attempt to ween him from such activities, since that might do more harm than good.

When he had learned to read at the age of four, he was permitted to browse the lower shelves of the household bookshelf. Soon after, he was found immersed in a top shelf book, his father's Analytical Geometry text from college. His parents thought he was just admiring the pretty pictures, another sign of a potential artist, but then the questions started.

"What are these words?"

"Right triangle."

"Why is a right triangle special?"

By sixth grade, Martin had developed a strange fascination with the number pi, and he programmed the family computer to print pages and pages of the number. His science fair project in his freshman year of high school was so far outside the realm of his teachers' understanding that they called in a professor from the local college to offer an opinion. Martin and the professor talked about the equations for nearly an hour. By the end of the week, Martin and his parents received a luncheon invitation from the college, and it was decided that Martin would continue with the usual high school courses, but take math courses at the college, full expenses paid.

Some other high school students might have been uncomfortable with the idea of leaving their classmates and mixing with students quite a bit older than themselves. Martin was a solitary individual, so it didn't really matter what crowd he was in at any moment. Fortunately, the practice of mathematics is easier for solitary individuals.

After high school graduation, he began at Berkeley. It was as an undergraduate at Berkeley that he gained his initial fame. Using his programming skill, some deep knowledge of an obscure lemma of Galois, and idle computing cycles on as many computers he could reach on the campus network, he found an optimum Golomb Ruler of length twenty seven.

In graduate school, he combined his dual interests in pure mathematics and computer science with a thesis on cryptography. It was not unexpected that Roos would earn his PhD in mathematics in record time, and that he would win the Fields Medal, the mathematics equivalent of the Nobel Prize. What was unexpected was his burn-out at the early age of twenty eight. He renounced all interest in mathematics and retired to a free-range chicken farm.

Since his worldly wants had always been small, he had saved a considerable amount of money, so the down payment on the chicken ranch hadn't been a problem. Tom, a college acquaintance, had helped him write a business plan to secure a business loan. The depth of the business plan was likely overkill for the rural bank with which he dealt. For Roos, the business plan served as more than just a means to extract money from a bank. It was an algorithm for an assured income, and he believed that strict execution of this algorithm would be essential for survival.

Roos had chosen chicken farming by an algorithm, also. He needed to get away from it all, so living at a rural location popped up as one logical consequence. Cranking through the numbers of the profitability and administrative demands of different types of small farms, Roos was left with a choice of organic produce or free-range chickens. Organic vegetables were seasonal, so they demanded too much work at certain times of the year. Chickens, however, were steady, year-round work, so he chose the chickens.

As he entered this new profession, Roos couldn't help but make the analogy between what he was doing and the plot of the 1960s television series, Green Acres, in which a New York City attorney moves to a farm to get away from it all. In that case, the man had a wife in tow, and the wife missed her former luxury lifestyle. The attorney knew little about farming, so his rural life was a mess. Roos hoped that the problems were just exaggerated to make for a good show, but he wasn't taking any chances.

Just as he had absorbed so much mathematics in his day, Roos did a careful study of chicken farming, and of the chickens, themselves. He knew that a small operation such as his must operate in a niche, so that's why he decided on a free range farm. He would produce not just free range chickens, but organic chickens. This meant that the chicken feed must be a hundred percent organic; that is, the grain must not have been genetically-modified, and it must have been grown without chemicals such as fertilizers or weed killers. In short, his chickens were luxury fare, so he could demand high prices in specialty store distribution.

Fortunately, free range chickens needed little tending, and an operation such as his could get by with just a couple of hired hands. They were always busy, fixing fences, spreading feed, hatching chicks, and doing weekly harvesting from the flock. Roos was happy that the harvesting just involved putting the chickens into cages for slaughter elsewhere. He was happy, also, that the rural men he hired had a very good work ethic, so the farm ran on automatic. The first year was an adventure, but after that it became more like a routine.

So here he was, paternally gazing at his chickens, oblivious to the forces that would thrust him into the greatest mathematical puzzle since Fermat's Last Theorem.

4 START-UP

Four weeks had passed since Tom and Rick's milestone meeting with the venture capitalists. The consultant had done his homework, and their two-man company was now twenty million dollars richer. The wire transfer came just days after the meeting, but along with the money they got another employee - an interim Chief Financial Officer "recommended" by their sponsors. They could hardly refuse. Besides, they had no idea how to find a CFO if they needed one.

Dan, the new guy, kept a low profile and seemed easy to deal with. He was clearly of the opinion that you had to spend money to make money, so he was no problem at all. To celebrate, they got new business cards with a logo Rick had developed from his college days. Tom, of course, was President, while Rick became Chief Technology Officer and Chief Architect. Dan decided he didn't need business cards at this point, and thus exercised his first fiduciary responsibility.

The first order of business was obtaining the raw computing power they needed. In the old days of computing, which were just a few decades prior, you would order a supercomputer, and then wait two years for it to be built and shipped. Nowadays, you could requisition the appropriate amount of raw computing power from a "cloud" service; or, you could assemble your own supercomputer in-house from readily available components. For security reasons, and because he needed to know precisely how his computer functioned, Rick decided to build his own in-house supercomputer. To that end, he ordered 8,192 off-the-shelf computers with individual computing power a step up from a standard desktop.

These computers, however, were not intended to work with keyboards, displays, mice or monitors. The money you would have spent on those peripherals was spent instead on high-speed optical network connections and extra memory. Some software running on each machine and one supervisory machine was used to "glue" all these computers into one, huge supercomputer. The basic programs for doing this were available free on the Internet, but it took someone with Rick's skill to make them sing.

Along with the computers would come their largest operating expense next to rent and salaries - electricity. The cost for running all these computers was about $75 an hour. This, coupled with the extra air conditioning needed to remove 500 kilowatts of heat, gave their CFO, Dan, something to fret about for the short term.

Rick had been amassing snippets of computer code for years, and he started gluing all the pieces together on a smaller, desktop computer. The most important piece of code was the one designed for getting the raw market data into a useful form. The raw data was piped in from the same type of service that provides real time quotations to the major brokerage houses. Rick's programs were designed to scan the data for trends based on an ever-changing base of rules derived from his evolutionary kernel program. Soon he was able to run small simulations to verify that they could make more money than they would lose. Eventually, when everything seemed to be working, the code would be ported to the supercomputer, and they would link up with a brokerage to execute actual trades.

One factor that worried both Tom and Rick was the idea that they were analyzing the market without taking into account the affect of their own trades. This sort of self referential idea had been bugging philosophers for at least a hundred years. It seemed to be implicit in quantum mechanics, also. Rick likened it to reentrant code in a computer program where a piece of code calls itself to do a calculation. As long as this happens only a few times, computers can handle it. It's only when the process does not complete, or your calculation doesn't converge within a set limit, that you have problems. Tom's hypothesis was that the market was so huge that their individual trades would be little more than noise in the system, and they would cause no ripples in the pond. Tom even did some simulations that showed his hypothesis was true, as long as they didn't get too greedy. Rick had even programmed a greed factor into his programs, and the factor was set rather low at the start. They would try to optimize performance as a function of greed when everything was running.

One condition of Tom and Rick's funding was a schedule and milestones. Although these were constraining, both admitted that they would keep them on track. It wasn't anything different than school assignments and grading. Rick called it a "chicken and egg problem." Did life mirror school, or was school designed to mimic life? Whatever the case, it kept them focused on their goal.

Deciding on the schedule had seemed to be as difficult as doing the scheduled work itself. Quite often in the process, Rick repeated a rule he had heard in graduate school, that all software projects took twice as long as expected, even when the rule was taken into account. It was best to create your best estimate, and then multiply by two for a safety factor, hoping that the factor would be less than two for some tasks to balance those greater than two for others. This is why spreadsheets were invented.

Renting a building, specifying the computers and getting them installed were some of the more routine tasks, but they took up a fair chunk of their money. Fortunately, these tasks simultaneously populated a large fraction of check boxes on their master plan. To the outside world, Rick and Tom were successful at this point, but only on paper.

In the weeks that followed, Tom spent most of his time reading some mathematically intense finance papers in an attempt to get further insight into the market forces they were attempting to turn to their advantage. Rick was running some test programs using real market data, but still no real money. He found that it was very rare when he would lose money, but the amount of money he was making was still quite small.

"Time to crank-up the greed," he thought, reasoning that there would be loses more often, but a lot more gain, a factor called scalability. What he found, instead, was that his system entered a mode in which there were loses for extended periods, followed by gains for other extended periods. His evolutionary algorithm needed tuning, since the ideal response was a continuum of gains and losses in which his program made money for whatever time, short or long, it was operating.

Rick had reached the mode that all computer programmers hate - Debugging. The problem with taking the bugs out of a program is that there are so many places where they could hide, just like bugs in the real world. It might be a case of counting from one instead of zero, or having a subroutine that's never run because it's only called on a condition that doesn't exist. There were software tools to aid in debugging, but Rick's preferred method of debugging involved staring at printed sheets of code with a red pencil in hand, looking for divine inspiration.

While the men were tending the shop, Connie and Mary were growing accustomed to their new lifestyle. Their new lives, so different from their former lives as the wives of struggling students, was like a paradise; but, it was a paradise with limits. Although Tom and Rick were pulling down regular salaries, they were not large by the standard of the day, and they were living in an expensive region. Still, it was nice to bypass the soup aisle while shopping at the local supermarket.

The big attraction to them was a huge shopping mall located very near to where they lived. The mall had a good mix of upscale and discount stores, and nice places for a leisurely lunch. They tried not to go there too often, but at least once each week was a necessity. On those trips there would first be a pass through those stores they couldn't afford, just for entertainment sake. After that, it was lunch and some earnest shopping in the more reasonably priced stores.

It was always good to have an instant critic while trying on clothes, and someone to encourage a purchase when the price seemed a little too high. It was on one of these excursions that Connie and Mary decided to overhaul their spouses' Casual Friday look. This was especially important, since Casual Friday at their operation was every day of the week, unless there was some off-site business meeting. The girl's scheme, which worked as planned, was that a simultaneous assault on two fronts couldn't be rebuffed.

The boys resisted at first, but they were weaned from their blue jeans to some yuppie-wear within a week. The blue jeans were kept safely in a drawer for occasional weekend yard work and home repairs.

As new home owners with modest salaries, Tom and Rick needed to do their own home maintenance and repairs. Visits to the nearest home goods store happened every weekend, and the wives would usually browse the garden shop while the men tried to make sense of the vast variety of tools and supplies.

Not surprisingly, Rick was a whiz at fixing broken door bells and light switches, and he didn't mind lawn mowing. Such mindless work as lawn mowing left his imagination free to explore solutions to the vexing problems of the previous week. Tom wasn't as technically skilled as Rick, but the simple recourse of do-it-yourself videos on the Internet proved sufficient in most cases.

Tom enjoyed his trips to the home goods store for another reason. As an economist, he was a student of the human condition, and it was interesting to observe the wide spectrum of people shopping there. There were professionals, such as Rick and himself, mixing with tradesmen speaking a plethora of languages. In a store such as this, the knowledge base of the tradesmen was far more useful than any training that Rick or he had received.

Tom remembered a humorous story in which visiting space aliens determined that the automobile was the dominant species on Earth. This automobile species was well tended by the slower and weaker symbiotic species that we call humans. He didn't think that space aliens would be that stupid, but they would surely lump everyone in the home goods store into the same category, fancy degrees being irrelevant.

5 BIG EAR

Tukey rounded the corner and got his first-of-the-day glimpse of the giant radio telescope antenna that beckoned him to work each day. It was a hundred meter marvel of engineering with a surface matched to a paraboloid to within a few millimeters. Such high precision was needed. If it were less, the antenna would be inefficient at focusing the short wavelength radio waves used for observation. With a structure of this size, some compromises were necessary. A structure so massive as to keep such a precise shape at any tilt angle and in high winds would not be able to support itself unless constructed from exotic and expensive materials. Instead, the lower surface of the structure was dotted with hundreds of position sensors and magnetic pistons that worked in concert with some complicated computer systems to push or pull the surface to bring it into shape.

Radio astronomy antennas had come a long way since 1929, when Karl Jansky quite accidentally built the first radio observatory at a Bell Laboratories testing station in Holmdel, New Jersey. His antenna was a fifteen meter directional wire array that looked more like a clothesline than the parabolic antennas in use today. Jansky was a physicist doing research into noise sources that interfered with radio communications. He found noise from lightning strikes, and another, steady hiss that changed in amplitude in a cycle just four minutes less than 24 hours. This is the period of a sidereal day, the time it takes for the stars to circle the earth, so this steady hiss came from the stars, and not the sun or earth. Jansky published his results, they were front-page news in the New York Times in 1933, and so he became the first radio astronomer.

It was surprising, but no flurry of research in radio astronomy began after Jansky's success. It could have been that astronomers were not interested in radio, and radio engineers were not interested in astronomy. It wasn't until 1937 that an amateur radio enthusiast, Grote Reber, designed a 30-foot parabolic dish antenna and built it in the backyard of his house in Wheaton, Illinois. Reber chose the paraboloid shape and a high radio frequency for the same reasons they are used today, higher angular resolution. Reber's antenna was made from a wood frame and iron sheeting. The weight of the iron sheeting alone was more than a thousand pounds, and the wooden framework brought the total weight to about two tons. The world's first parabolic radio telescope antenna cost Reber about $4,000. In today's terms, that's about $50,000. Reber was a very dedicated amateur astronomer! For all this effort, his first observations were unsuccessful. It was only after migrating to lower frequencies that he was able to produce the first radio maps of the sky. Reber was the only radio astronomer until the end of World War II.

The advent of radar in World War II led to a surplus of radio equipment and a cadre of engineers and scientists who had experience in building and improving the wartime radar systems. Among these was Bernard Lovell, an English physicist who used radar to study meteors. Lovell championed the construction, at Jodrell Bank, of a monstrous 250 foot fully steerable paraboloid antenna that was completed in 1957. It was this observatory that verified the presence of the Russian rocket that launched the first artificial earth satellite, Sputnik I, on October 4, 1957. Sputnik launched the world into the Space Age, and it also caused a surge of interest in radio astronomy.

In the United States, John Kraus designed a radio telescope called the Big Ear. He and a group of students at Ohio State University built this radio telescope in the late 1950's for about $250,000. Comparing this with Reber's $4,000 telescope, you can see a definite trend in the cost of equipment! The government also took notice and established the National Radio Astronomy Observatory in remote Green Bank, West Virginia, where it would be shielded from most man-made interference. Of course, those were the days before cell phones and direct broadcast satellites. The NRAO's premier instrument was a 300 foot parabolic antenna built in 1962. It collapsed in 1988, a victim of metal fatigue, only to be replaced by the more versatile 100 meter Green Bank Telescope. Perhaps because of national pride, the Green Bank Telescope is just slightly larger than another 100 meter telescope at Effelsberg, Germany.

Blaise Tukey was born in 1969, the year of the first moon landing, but by the time he had entered school, the initial glow of man in space had vanished. Seeing man's conquest of the moon as a *fait d'accompli*, the public rarely tuned their TVs to watch the then frequent moon jaunts, despite moon buggies and live color television from the moon. The last man walked on the moon in December, 1972. But the Viking, Mariner, and Pioneer planetary probes returned spectacular photographs of most of the planets shortly thereafter, and Blaise was hooked on astronomy.

He immersed himself in the astronomy section of the local library, first intent on viewing all the pretty pictures, then memorizing the constellations and the names of the planets and their satellites. He built his first optical telescope in the sixth grade, polishing and testing his twelve-inch mirror using techniques that had changed little in hundreds of years. Earning his stripes through many a cold night, he gradually integrated himself into the local amateur astronomy club where he learned even more. Soon he realized that although his club mates had a lot of technical knowledge, they possessed only limited knowledge of real astronomy as practiced in the universities. At that point he stopped attending meetings and went off to do his own thing.

Then he discovered radio astronomy. His local library had just two books on the subject, one of which was an ancient amateur's guide with vacuum tube circuitry. The other was an illustrated history, almost current to the present day, with wonderful photographs of antennas and control rooms full of electronic equipment. He was hooked, but he realized that the pursuit of radio astronomy required as much knowledge of radio electronics as astronomy.

He had noticed that a neighbor down the street had a very large antenna on a tower alongside his house that obviously wasn't used for television. His neighbor was an amateur radio operator, or Ham, as they called themselves, and Blaise built up his courage to knock on the door. He was glad he did! He got a grand tour of his neighbor's radio shack and talked to New Zealand via radio. He got an armful of books to bring home, and a promise that he would be tutored enough to become a licensed radio operator. It appeared that the ham radio hobby propagated in a master-apprentice fashion.

The main hurdle to getting his amateur radio license didn't involve learning electronics and the necessary Rules and Regulations. It was learning Morse Code. This code, an extension of the rudimentary code invented by Samuel F. B. Morse in 1836, was a hundred and fifty years old, but it was a requirement for a radio license since it could be used to transmit messages with primitive equipment in emergencies. Tukey devoted as many hours to learning the code and studying radio electronics as he had making his optical telescope mirror. All this study paid off when he passed the examination for an advanced class of license that allowed him to use nearly every frequency reserved for radio amateurs.

Tukey's father was a carpenter. His family didn't have the money to fund a well equipped radio shack like his neighbor's, but that didn't matter. Tukey was more interested in building radio receivers than transmitters. The small amount of money he earned doing odd electrical jobs and painting houses, amplified by a steady stream of surplus equipment he obtained from other Hams, allowed him to build sensitive receivers spanning all useful frequencies. In high school, he placed first in a science fair with a radio telescope assembled from one of his receivers, an antenna made from copper plumbing pipe, and a chart recorder. Like Jansky, he was able to show charts of the increased radio intensity at the galactic center, a radio source known as Sagittarius A.

After high school, his course was clear. He majored in Physics and Astronomy, and at the same time attended electrical engineering courses most appropriate to his career path. With such an excellent background, he was quickly accepted into graduate school to apprentice under a Professor of Astronomy whose own radio credentials extended back to the early years of the 1950s. His thesis, on fast amplitude fluctuations in pulsars, brought him to the forefront of his profession where he could have happily remained doing the almost routine things that other astronomers did. But there was a Siren call for him to devote himself to SETI, the search for extraterrestrial intelligence.

When Tukey was still more than a mile from the parking lot, he turned off his car radio. He also checked that his cell phone was really off, not just in standby mode, and he stashed it in the map compartment. These were precautions required of all employees to prevent spurious interference. The man on the street didn't realize that radios are also unintentional transmitters, and that cell phones, when not off, are transmitting their presence to the nearest tower. He did keep his PDA in his pocket. Like all other computer equipment in the observatory, it was Tempest rated; in other words, it was certified not to interfere.

The main antenna was casting its long morning shadow across the plain. The shadow didn't quite reach the control center, since the antenna was placed at a suitable distance to isolate it from other sources of interference caused by human presence. Humans themselves emit radio waves, but fortunately the emission only becomes detectable above 100 Ghz, a hundred times cell phone frequency, and almost a factor of ten higher than the highest observation frequency. That there was just one, huge antenna and not many smaller antennas, was significant. Although it was possible to synthesize the equivalent of a large antenna from many small antennas, the synthesis was not ideal. Instead of just concentrating reception in one direction, there was also spurious reception in other directions. Changing reception frequency for one antenna was also much easier.

It was now about 10:00 AM. Optical astronomers worked by night and tried to sleep by day, but radio astronomers kept time by whatever objects were visible in their hemisphere, any time of the day or night. Tukey flashed his ID badge at the security desk and clipped it to his open collar. Security had gotten increasingly tight over the years, since there was one lunatic fringe after another knocking on the door without an appointment. The huge antenna, if nothing else, was an icon of UFOs, extraterrestrials, and their ilk. He fetched his mail from the hallway slot and decided to bypass his office to go directly to the control room.

Brian was not in the control room, but that was no real surprise. The technical staff kept things on an even keel, and most things were highly automated. Scientists were needed at the occasional decision points, but impromptu decisions were rare. Most things were carefully planned because of the huge hourly cost of using the instrument. Tukey turned down the skylighted corridor towards the office wing. He found Brian, not in his office, but in the nearby office of Marianne, one of the graduate students. They were studying a computer screen when Tukey knocked on the open door.

"So, what have we got?"

"Well... things aren't exactly clear, yet. We're still crunching the high resolution data from this morning's transit, but we may have missed the source, if there is a source, on the first scan. It's like the Heisenberg Uncertainty Principle for astronomers - The closer you look, the more you need to know where to look. You can resolve the hairs on the head of a fruit fly with a high powered telescope, but the fruit fly had better hold still. We'll just keep stepping until we see a strong signal, or admit we were fooled."

Tukey directed his attention to Marianne.

"It's nice having someone to share our SETI fantasies with, but how's the *real* research program?"

"Well, if you can call the Krumhansl Survey, research, it's going well. I think it's more like stamp collecting, or one of those Victorian bugs-pinned-to-a-board sort of thing. It's the things that I help Brian with that keep me going."

That was one thing Tukey enjoyed about Marianne. She wasn't yet a PhD but she talked like one, speaking her mind and letting the other side interpret as they may. Tukey was like that as a graduate student, and he was certain Brian had been the same. Was this a learned response for scientists, or were scientists selected from a group possessing these traits? Perhaps there was a sort of kin selection effect at work in his group, or at work in the profession in general.

"Well, I'm sure it works both ways. I mean, would Brian be interested in such an unrewarding enterprise if he didn't think other people were interested? What can I say, except keep up the good work, both of you."

His minimal professorial duties of the day completed, Tukey retired to his office to check his e-mails and budget sheets. The next transit sweep wouldn't occur until tomorrow, no matter what timeline was laid into his project plan. The rotation of the earth would stop for no man, research director, MBA, or otherwise. Tukey settled down for another bout with the journals, and maybe a nap after lunch.

6 FINANCIAL MODELS

Electronic calculators had always been central to finance. NCR Corporation, a pioneer in minicomputers and parallel processing supercomputers, had been called National Cash Register. In 1914, an NCR executive, Thomas J. Watson, Sr., joined another company, The Computing-Tabulating-Recording Company, and changed its name to International Business Machines Corporation; or, more commonly, IBM. Until the early sixties, computers and calculators were just number-crunchers and an easy way to print pay checks.

Economics had always been a strange cousin to finance. Economics deals with equations, not just numbers. Finance involves things that can be done on the simplest calculator and maybe the occasional exponentiation for mortgage calculations, whereas economic equations often involve calculus and other higher mathematics. But it's more than math. Its model building and theory forming is more like physics than keeping shop. Scientists who ventured into the financial markets were often called *rocket scientists*, and Rick wore this appellation proudly. He would caution, however, that it's much nicer to ride the rocket on the way up than down.

One of the greatest advances in computation economics was the Black-Scholes derivative pricing equation derived from a model that attempted to equalize the risk and reward aspects of investing. From this beginning, an entire field developed to apply these models to niche market areas. The object of these models was to transform rocket science into money. Tom and Rick were attempting to do the same, but they had staked out an undeveloped area. Natural phenomena based on random events typically followed the Gaussian Distribution, commonly called the Bell Curve. Human individuals think their lives are purposeful, and anything but random; but, in the aggregate, humanity as a whole is a random ensemble. As Rick would say, "People are like gas." Individual gas molecules in a balloon do their own thing, traveling fast, slow, in between, this way, or that, but in the end they average out to give a constant pressure that keeps the balloon inflated.

Fortunately for Rick and Tom, the gas model is not exactly true when it came to stock trades. Some people choose not to sell, even when their stock dives to a tenth of its value. Others hedge their potential losses by buying more stock in the same company, thereby reducing the average acquisition cost of the stock. Likewise, some people are too suspicious to sign onto the next sure thing. The result of this is that the Bell Curve has fat tails. When the wives, then just girlfriends, had first heard the men discussing "fat tails," they were insulted. The economic explanation seemed like a coy cover-up, but after a while they became believers, and the men married them, fat tails notwithstanding.

Rick and Tom were attempting to analyze the Bell Curve of stock trades in real time, and to capitalize on human psychology. As Rick would say, they were going to make money in the tails. One advantage of this approach is that timing wasn't that critical. Other computer schemes worked in an arbitrage regime in which every microsecond was important. Rick and Tom were working at the seconds, perhaps minutes, level.

Rick's fat tail analysis was based on a self-improving model, an evolutionary program that would optimize itself in response to its wins and losses. The wins and losses would be just paper wins and paper losses at first, while he got the bugs out of his program. Such programming required the computation power of a supercomputer.

Tom and Rick had many options in specifying their supercomputer. As Rick said repeatedly, when you look beyond a certain threshold, everything was scalable. What that meant, in effect, was that a small computer would make a small amount of money, and a large computer would make a large amount of money. After some mathematical modeling, Rick calculated the minimum computer that would give them a break-even operation, just covering their operating expenses, and they arbitrarily doubled it. The doubling was either a safety factor, or an anticipation of a 100% profitability. It was probably somewhere in between.

There was the business plan, and then there was Rick's plan - All the tasks he needed to complete to get their computer process working. Computing is logical, and so was Rick's plan. The master program was split into many separate and independently operating units. He could program and debug each of these modules separately, and when they were all working, he would glue them together.

There was a module that received the raw stock ticker data, another that filtered it and put it into a convenient format for computations, and several others that handled the computations. The computation programs were different than those of a typical mathematics program. Although their calculations were always precise, the calculations that were being done were not decided by Rick. The computer made its own decisions. When the program was started, it made trades at random. By keeping score of the winners and losers, it was able to train itself to pick winners.

Rick's major occupation after the computer was operational would be to monitor its performance. He had developed a metric, called a fitness function, that told him how well the system was performing. The computer could unload a stock when its value went up a penny – pat itself on the back, or whatever doubles as that body part on a computer, since it was making money – and move on to other things. Rick's fitness function would see that after that penny's profit, the stock zoomed upwards several dollars, and it would caution Rick to suitably chastise the computer. He would do this by analyzing and modifying the code.

But the time for planning was over. Soon the computers would arrive, and Rick would set his programs loose on the real world.

At the same time, Tom was following up on whether there might be a use in their scheme for the latest buzz word in finance, Tsallis entropy. Since their model financial system was like a bag of gas, anything related to thermodynamics might be useful, and entropy was a big part of thermodynamics. Classical entropy was a useful concept, since it set a level playing field for transactions. There were no preferred traders and no certain outcomes.

Classical thermodynamics had the problem that it could only be applied in equilibrium systems in which a trade in one direction was identical to the reverse trade; that is, the process was reversible. In the case where you were always trying to make money in a trade, you were introducing some irreversibility. If the profit taken was small compared to the investment, this wasn't that much of a problem. If you went for higher profits all the while, there might be some catch in the equations that would get you in the end.

Tsallis entropy was a new way of looking at entropy in non-equilibrium systems. Tom saw that its application improved some tasks in the physical sciences, such as signal and image enhancement. Fortunately, the mathematics wasn't that intense, so Tom was enjoying his research. What made it all the nicer is that he was getting paid to do this homework assignment.

7 NUMBER CRUNCHING

"Ah, there's nothing like the smell of computers in the morning!"

Despite the air-handling equipment, hot computers still emitted a smell not unlike the new car smell of yesteryear. This was the smell of volatile organic chemicals mixed with an equal part of motor oil from the cooling fans. Fortunately, from a respiratory health standpoint, the smell faded after a few months, but it never quite disappeared.

Rick rounded the corner and noticed, as he passed the offices, that, once again, he was the first to arrive at work. Most programmers worked a strange shift schedule of roughly noon to midnight, and perhaps later. Such a schedule had been imprinted on them through long nights of impossible homework assignments. If you asked a programmer to attend a nine AM meeting, he would usually respond that he normally didn't stay up that late. Rick had somehow outgrown this right of programmer passage and replaced it with the joy of working in seclusion.

Things had gone smoothly in the hardware installation phase. The cabinets that would house the separate computers of their cluster arrived first, and they were wired for power by professional electricians. Since there would be 128 computers in each cabinet, they needed 64 cabinets in all for their 8,192 computer cluster. Rick had hoped to arrange these in an eight-by-eight array (programmers are hung-up about a powers of two thing), but their space wouldn't allow this. They ended up with four rows of fourteen, and a bastard row of eight. Rick still hadn't gotten over his irrational dislike for this configuration, but the short row allowed space for the central operating desk with its four large monitors, multiple keyboards and master computer. This area was isolated as a room-in-a-room to shield the operators from the steady noise of the cooling fans. They had investigated liquid cooled computers, but liquid cooling was too expensive, and they had enough problems without needing to mop floors after the inevitable leaks.

After the cabinets were wired, it was time to install the computers. These were plugged into small boxes just a few inches high that stacked one atop the other in the cabinets. The combined height of thirty-two of these, plus a large space at the bottom for cabinet cooling fans, completely filled each six-foot cabinet. A herd of students from a local college was hired for a weekend to uncrate everything and install the computers in the cabinets. Bolting the computers to the cabinets was the easy part. The hard part was the cabling. Fiber optic cables interconnected computers in each cabinet to a router, and then each router was connected to the master computer by fiber optic cables in overhead trays. The master computer, which was also called the supervisory unit, fed instructions and data to the computer array and received their results.

When all 8,192 computers looked as if they were working, Rick pummeled them with some diagnostic programs he had written in anticipation of that moment. More than eight thousand computers, each with billions of tiny transistors wired to execute his every whim! That's more than a trillion transistors doing billions of operations per second. Rick remembered that ENIAC, the first large scale general purpose computer, had less than 20,000 vacuum tubes. It could do just a few hundred mathematical operations every second.

Computers are generally reliable, once they've been shown to work for the first time. During Rick's debugging phase, a few were flagged for errors, most of which were corrected by tightening the existing fiberoptic connection, or replacing a cable. One computer seemed to be DOA, until Rick noticed that the switch on the power supply was set for a higher voltage they weren't using. Having every computer operational out of the box wasn't as unlikely as you would think. A "Six Sigma" manufacturer ships fewer than four defective pieces of equipment for every million that it sells.

There were two types of "glue" that connected these 8,192 individual computers together to make a single supercomputer. There was the fiberoptic network cabling, but there was also the supervisory software that made certain each unit had enough data to do something useful at all times. Much of the software concerned the communication of intermediary results between units. Such data were routed through near neighbors when possible, but often the supervisory computer needed to handle the transfer. That's why that computer was faster than the others.

In order for software to function as quickly as possible, Rick tried to write many software routines in machine language, a code nearly unreadable by humans, but very easily understood by machine. Other, less time-sensitive programs, were written in high level languages that made coding easier. In the end, these high level programs were automatically converted into machine language, since that was really the only language that a computer understands.

Rick was in his own element, and he was on a roll. Somewhere in the back of his mind was the idea that he was doing this to make some money. In reality, he was doing it because he enjoyed it.

One thing different between Rick and Tom was their preferred working hours. Whereas Rick would arrive very early in the day, thereby avoiding the heaviest rush hour traffic, Tom preferred banker's hours. Tom was able to avoid some traffic during his morning commute, but his hours were set more by the people he needed to interact with during the day than his own preference. Since he handled the business side of things, he needed to keep the same, regular office hours that others kept. Rick's interaction was with his machines, which worked 24/7 and could care less with whom they interacted, and when.

Still, there were quite a few overlap hours between the two; and, when Tom wasn't at some business lunch, they would have lunch together, usually experimenting with different cuisines as an adventure. They had really enjoyed a nearby Afghan restaurant, and they were sorry when it had closed. Even without that, they was no loss of variety as they jumped between Thai, Chinese, Indian, Mexican, the many burger joints, and several other budget options, at the flip of a coin.

That day, Rick and Tom opted for something somewhat different. They went to a vegetarian restaurant. It was not because they felt that their bodies needed a purge, but Tom had heard from a business contact that the food was surprisingly superb. The restaurant was located in a storefront at a strip mall in a marginalized region between a decaying neighborhood and a neighborhood undergoing gentrification. The new money class from Rick and Tom's generation needed places to live, although few like they could afford a suburban home; and, they liked their restaurants nearby.

Fortunately, this was a vegetarian place that didn't think that fried food was poison, and that's likely one reason why most menu items had a great taste. Tom remembered his mother's breaded and fried cauliflower from his childhood, so an obvious first choice from the menu was a fried tofu and cauliflower medley. Rick chose the eggplant, and they ordered a starting salad and buns stuffed with potatoes and assorted vegetables. For drinks they ordered mango milkshakes.

As they dug into their salads, Tom continued their luncheon conversation with a work-related topic.

"Rick, I've been reading about quantum computers, and it's got me a little worried. If these things are so much faster than supercomputers, is some upstart company going to use one of those to do what we're doing, and eat our lunch?"

Rick waited a second to swallow his food, and then he responded.

"Well, unless what you were reading was just a rehash of some press release, you would also have read that many people aren't sure whether quantum computers will be any faster that the usual type; or, maybe, they might just be useful for a very specific task. The ones that I'm familiar with, quantum annealers, would be good for finding the best route to Grandma's house, but not much more."

"Yeah, that sounds like what I read, but scientists are doing a lot of work on these things. Maybe there will be some breakthroughs."

"Well, the biggest problem might be a fundamental limitation. The "qubits," the essential computing elements of quantum computers, need to be in a combined state called a superposition during quantum computation. Right now, the time period for superposition seems to be too short. Maybe longer superpositions are fundamentally impossible."

"So, the important question is, what's the shortest time you estimate it will take, fundamental questions aside, for quantum computers to become practical?"

"Ten to fifteen years, as a very rosy prediction; namely, long after we've cashed out of our venture."

"That's a relief."

"As if that was our only worry..."

Tom laughed, and then it was back to his salad.

8 SETI

Tukey arrived home at about eight. He always left work at a late hour, since traffic was too heavy around the 3-7 rush. When his observations ended in the rush hour time frame, he would often work late just to keep his sanity. He instinctively flicked on the television to no particular channel and walk into the kitchen for some food. He didn't really care what was on (a thousand channels, and nothing to watch, as the saying goes), and since he got all his news from the Internet, he was sure he wasn't missing anything. He zipped open a bag of tuna, plopped a fork into it, and retreated to his favorite chair. The tuna went down quickly. He eased back into the chair, and he succumbed to his usual post-dinner stupor. In his half waking state, he reflected on the events of the day and reflected on his profession.

There were always those who would push the technology of their day to contact extraterrestrials. It all started with Mars, and Giovanni Schiaparelli, the Italian astronomer who found the "canals" of Mars in 1877.

Schiaparelli thought that these were just geological features, but others believed that they were waterways. Soon, there were maps with many canals criss-crossing Mars, and the land around these even seemed to change with the season, as if vegetation were blooming. An American, Percival Lowell, was fascinated with the idea that these markings could be a sign of a Martian civilization. Suddenly the Martian canals were a feature of a desert landscape, designed by the possibly extinct Martians to bring water from the poles to equatorial oases of vegetation. He wrote a popular book, *Mars as the Abode of Life*, expressing his vision.

SETI had become a true scientific discipline in 1960 when Frank Drake pointed the Green Bank, West Virginia, telescope at two nearby stars, Tau-Ceti and Epsilon-Eridani, and listened. It would have been embarrassing for the human race if someone were talking, we had the capability to listen, but didn't. Drake listened for two weeks, and then the telescope was relinquished for more important things. As if hearing from beings from another planet wasn't important!

SETI seemed to languish for many years thereafter, although a young graduate student, Jocelyn Bell, made an accidental discovery at Cambridge University in England that reminded everyone about the possibility of detection. Her faculty adviser, Anthony Hewish, had encouraged her to make measurements at a rapid rate as a way to find quasars. Bell's radio telescope looked nothing like Tukey's instrument. It was composed of thousands of wires connected by miles of cables in a way to boost reception. The frequency was low by today's standards, more like an FM radio station than a broadcast satellite. In 1967, Bell found a signal she couldn't explain that she described as a bit of "scruff" on her chart traces. The radio signal was periodic, a feature never before seen in radio observations. There was, of course, the obvious search to pin it down to something terrestrial, such as satellites, or radar waves bounced off the moon. But the signal kept sidereal time, star time, and not earth time, and it had features that identified it as coming from outside the solar system.

The Cambridge team jokingly referred to the signal as "LGM," an abbreviation for Little Green Men. There was the possibility that they had discovered signals from an alien civilization, but they had actually discovered pulsars, rapidly spinning neutron stars that give periodic signals because of their rapid spin rate. Hewish eventually received the Nobel prize for this discovery. Bell didn't, a fact that raised the hackles of many a feminist. There has always been the tradition that the student works under the direction of the professor, the professor gets the glory, and the student is blamed for the mistakes.

Ten years passed before the next potential SETI event, the "WOW" signal at Ohio State University in Columbus, Ohio. Astronomers commissioned a radio telescope called the "Big Ear" there in 1973. On August 15, 1977, when the "Big Ear" was pointed at the constellation Sagittarius, there was a signal that jumped off the charts. It was thirty times stronger than the background noise, and it had a characteristic that indicated that it came from a source outside the solar system that was scanned by the antenna beam. The operator on that night circled the reading, labeled it "WOW," and a SETI legend was born. Follow-up studies, including one with the sensitive Very Large Array telescope in New Mexico in the mid-1990s, did not recover the WOW signal.

SETI research faced some fundamental problems. There was the chicken and egg problem: what if everyone was listening but no one was talking? The cost of listening was small. All you needed was a radio telescope, already in place because of conventional observation programs, and a little electronics. Talking, however, required power, and lots of it. Megawatts of power.

There was also the problem of choosing the right frequency, although arguments for the "water hole," a frequency region near a prominent hydrogen marker signal, were persuasive. Hydrogen is the most abundant element in the universe, and this frequency region is also a minimum in the galactic noise spectrum. But a further problem was whether radio was the proper medium at all. A hundred years ago, Earthlings would have been looking for smoke signals. There was even a proposal at the turn of the last century to set fire to large swaths of forest, in the shape of geometrical objects, to signal Mars of our existence. Even now, there was talk that optical signaling, using lasers, was the proper medium. A hundred years from now it might be obvious to use positron beams, or something like that. Perhaps the alien technologies are so far advanced that we would have no idea how to communicate with them.

Then there was the problem of what to do if a SETI signal were actually discovered, although this was a problem Tukey would gladly face. There was a set of SETI protocols, endorsed by all astronomers, that governed what should happen if a candidate signal were received. The first, and most obvious, requirement was a verification that your signal was actually a candidate, and not some natural or man-made interference. If a true candidate still existed, the next step was to have it confirmed by your colleagues through independent observations and ensure that a continuous reception is obtained. It is only at that point that the discoverer informs his government, the United Nations, relevant organizations, and the general public. The discoverer has the privilege of making the public announcement. After that, the usual rules of science apply; namely, get the data and share the data, usually through conferences and publications. There are also some technical issues, like protecting the frequencies from interference. Lastly, no response would be sent before much consultation.

As in other scientific disciplines, colleagues exchanged their most recent results and theories at conferences. Technology may have given us email, teleconferencing, Internet message boards, and cellphones, but there was still nothing like meeting and talking face-to-face. Although the attendees were supposedly there to hear the formal presentations, the real action happened in the back corridor and coffee room discussions. There, preliminary data were divulged and collaborations hatched.

That year's International SETI Conference was conveniently scheduled for San Diego. Its proximity made travel very convenient for Tukey and Brian, and their budget had an ample allocation for travel. They prepared an abstract on a signal processing technique Brian was developing, and Tukey submitted the abstract to the conference committee. As Brian freely admitted, Tukey's name as an author assured acceptance. There was the Old Man's reputation, but there was also the careful oversight that Tukey brought to each research project. Eventually, the results of the Kumhansl Milliarcsecond Survey would be presented at a conference, but not at the SETI Conference, followed by a publication in which Brian's name would be just one of many.

Brian's computer skills had been a definite asset to Tukey's SETI project. Their main piece of hardware, the channelizer, was state-of-the-art, but it was useless without proper software control. The radio telescope grabbed a huge chunk of the frequency spectrum at one time, but any faint signals would be lost in the background noise of radio wave photons at each little frequency talking at the same time.

The channelizer sliced the broadband signal into tiny pieces, and it was able to see whether anything interesting was happening in any particular piece. Conceptually, this was easy; but, practically, there were some details that needed to be handled. The most important of these was the fact that the Earth was a moving platform. Earth's rotation on its axis and its revolution around the Sun caused frequencies to shift slightly. Since the radio signals passed through Earth's ionosphere, there was a scintillation, like the twinkling of a star. Correction was needed for all this, and also some corrections for subtle flaws in the telescope's electronics.

Brian had become an expert on these corrections as they applied to their telescope; and, as a good programmer, he had written his codes in a general enough fashion that they could be used at most other telescopes. This was the work that Brian would present at the conference.

9 CONFERENCE

The International SETI Conference was in San Diego that year, a short jaunt for Blaise and Brian, but they would have attended anywhere. It was made all that easier since Blaise was an expenses-paid invited speaker, Brian got a student discount, and everything not already paid was charged to their research contract. The last point was important, since it meant that Brian could stay in the semi-luxurious conference hotel and not in some distant, budget motel. Also, there was a tradition that students would tag along with the professionals for dinner, and the professionals would pick-up the tab, dividing it by N, where N was the number of non-students at the table.

The ISC was a true international conference, with other recent international venues in Leningrad, Cape Town, Trieste, Cambridge UK, Rio de Janeiro, and Vancouver. The United States was once the most popular venue, since the political climate had been amenable to short visits from scientists, and expenses were most reasonable. All that changed after September 11, 2001, but there were still too many scientists in the US to ignore that venue completely. There were some new precautions. Scientists who worked in the same building as facilities doing materials research were advised to buy new shoes for the trip, lest they trigger a false alarm from rare metal contamination at the airport sensors.

The US-based ISC conferences had been held in Boston, Chicago, Dallas, Orlando and San Diego, all locations conveniently near large research universities from which professors would volunteer for a local committee to handle most of the conference preparations. The first ISC was held in New York, but the conference committee vowed never to return, since it was far too expensive for a scientist's budget. Several dollars for a self-service cup of coffee from an urn was one thing, but the complaints from the exhibitors are what forced its hand. Exhibitors were the financial blood of the conference, since they paid thousands of dollars each for a small booth from which they hoped to generate sales of their latest computers, software, or radio frequency gizmos. In New York, they were forced by union rules to pay exorbitant fees to have their card tables unfolded and their curtains hung. The vendors threatened never to return to the conference if it were held again in New York.

There were other ways that the ISC committee kept the attendance fee for the conference low. The roster of invited speakers was skewed towards the Third World with little objection from the rest of the scientists. All of the First World scientists could tap into their research budgets for expenses. Room rates were kept low by booking hotels at off-season rates. So it was August in Orlando, or traveling on Easter, or the Sunday after Thanksgiving, to a conference scheduled for the next week. Sometimes the wives and children would tag along, but scientists generally traveled alone. A scientist with wife and family in tow did not experience the dinner discussions with colleagues where the leading edge of science was discussed.

The first morning was devoted to registration and a plenary session. Blaise and Brian parted company almost immediately upon arrival, since each had his own agenda. Blaise needed to visit with old friends and learn the subtext of the formal papers, and Brian needed to network, to line up his first permanent position. They did link up at lunch, when they, and three scientists from Slovakia, found an Indian restaurant with a nice menu and very reasonable prices. Marta, the most senior of the Slovakian delegation, a rotund and jovial person, said as they entered the restaurant, "Two Americans and three Slovakians went into an Indian restaurant... Sounds like the start of a very good joke!"

Blaise was a Chicken Tika Masala man, but one of the Slovakians was a vegetarian. They had selected the Indian restaurant in deference to him. While waiting for their meals, the conversation turned to the plenary address, with Brian leading.

"If I hear the Fermi paradox again, I'll just barf! I thought Tony was one of us!"

"Well, a lot of what he said was for the benefit of the Press. The Common Man thinks we pursue science in a logical, linear fashion, although we all know that's not true. Our public face is the historical-evolutionary perspective, so Fermi gets in there at the start, but we build from there."

"So we fake it for the non-scientists?"

“Well,” said Marta, “I wouldn’t put it that way, but every profession has to keep a polished face for the people who have the money. In Slovakia, as I imagine is also the case for the US, we have some government funding, but our field relies mostly on support from private foundations. Sure, the government provides the telescope for the real astronomers. Fortunately, we get to piggy-back on those since we pay our own salaries through private grants, and we can fund the odd accessory instruments we need.”

Blaise joined in.

“Most importantly, we scientists must be true to our profession. We can't ignore something like the Fermi Paradox just because it's bad for business. There's a lot to consider. Fermi, who was always adept at back-of-the-envelope calculations, realized that Earth should have had some contact with extraterrestrials, if they existed. He realized that the Sun is a Population I star, a youngster in the galaxy, and there are billions of stars much older than the Sun. If the evolution of life was possible on just a few of these, we should have had visitors by now. And, they needn't be biological creatures. They could just be robotic probes. Fermi thought that it would take just a few million years for any intelligent civilization to take over the entire galaxy. He summarized the lack of such evidence with the simple statement, 'Where is everybody?'"

"The trouble is that everyone takes that to prove that there are no extraterrestrials."

"The Rare Earth Hypothesis."

"Right."

"Well, it just might mean that they purposely avoid us because they don't want to destroy our culture. You know the examples, here on Earth, where modern men demolished the cultures of the primitive people thy met, and they decimated their populations as a side-effect. You remember the Star Trek *Prime Directive* business."

"Well, in that case, there's still no use in our listening, since they wouldn't be beaming signals our way."

Brian, who was a Star Trek fan, found a way to enter the conversation.

"It would also need to be a pretty tight ship. I remember Kirk and Picard used to violate the *Prime Directive* all the time."

“Quite right,” said Tukey. "It might be that just the really old civilizations, the ones that Fermi thought should colonize the galaxy, are extinct. There are enough natural hazards out there, gamma ray bursters, supernovae, and things we haven't yet discovered, that could wipe-out biological life in a fairly large radius. But, we don't need the Fermi Ancients for communication, just people like us."

"Not just natural disasters, but suicide. Perhaps all intelligent life is destined to destroy itself, either by war, or by some scientific accident. Maybe that next new particle accelerator will create a planet-gobbling black hole."

"There's one other important factor that's close to the hearts of this group – Economics. Interstellar travel must be expensive, and if you look at it from a per capita basis, that's a lot of money to spend on moving a person from one place to another place that's not that much different. Economics is an important part of our own programs. We listen, but we don't broadcast. Maybe everyone is just listening."

Brian interjected, “I guess the *per capita* cost would be lower if the aliens have more than one head.”

Everyone laughed, and Tukey continued.

"While we're on the topic of listening, maybe we're doing it all wrong. We've been searching for just a few decades, and we're hung-up on using radio. There are some good arguments, now, for optical SETI, but only because we discovered the laser. What might be discovered a hundred years hence that will seem more like an optimum communication technology, at least until a newer one comes along. Maybe there's a way to propagate waves in the zero-point field, or something like that. The time window for radio communication between stars might just be a few decades."

"There's also the idea that biological intelligence is just a transitional state to machine life. A self aware computer might be just a few decades away. After that, it's exponential intelligence, since computers will start designing themselves. It's been said that intelligent robots may be our last invention. Humans might end up as noise in the system, as something not worth worrying about."

"I don't think you can ever have intelligence without curiosity. Whatever it is that's intelligence will realize that there's no joy in just surviving from minute to minute. You always need something to do, although I don't know exactly how that would play out in a universe populated by robots."

"Do you think that emotions, such as compassion, are a part of intelligence?"

"I hope that's true, otherwise biological lifeforms would be just another raw material for a robot culture. Fortunately, they wouldn't eat us, and they'd be more interested in a good chunk of silicon than a messy blob of organic chemicals."

"Would there be individuals in a machine society?"

"That's an interesting question. They would certainly do R&D of sorts, and subsequent machines would be better, different. It would be like evolution, and there's also the possibility of the unexpected, like when a cosmic ray changes some memory bit so one robot behaves just a little differently than planned."

"Wouldn't there be self-repair systems, like our immune system, to handle that?"

"The self-repair system is also subject to cosmic rays, and if it had its own self repair system, the same would be true."

"There's probably another consequence of robots, rather than biological organisms, in control. Humans have a way of expanding to fill their environment, thus Fermi's idea of a wave of colonization taking over the galaxy. Perhaps robots see no need for useless reproduction. That would explain the absence of the colonization wave that Fermi wondered about."

I'm uncomfortable with this whole idea of being replaced by machines," said Brian. "I prefer my extraterrestrials to be organic and green."

Marta, who was the lead joker at the table, had a quick reply.

"You mean that they don't eat genetically-modified organisms, and their spaceships are energy-efficient?"

Everyone laughed as the servers produced their dinner plates to many "oohs" and "ahs." The pace of the conversation slowed a little as everyone started to eat.

After the first few minutes of obligatory face-filling, Blaise continued the conversation.

"There's another school of thought that Earth was visited by aliens in the dark past, but they didn't find anything too interesting, so they left rather quickly. In any event, they may have intentionally, or unintentionally left something behind - Some artifact that's somewhere still to be discovered. That's sort of the plot of the 2001 movie. In that case, the artifact was left on the Moon, presumably since it would only be discovered when humans were technologically capable of understanding it."

Brian interjected.

"According to the panspermia hypothesis, *we're* the artifact that was left behind. Some extraterrestrials left their garbage on Earth, and those living cells evolved to everything we have today."

"That's a nice story," replied one of the Slovakians, "but if Earth's conditions were favorable enough for such life to thrive, it would also have been favorable for abiogenesis long before that time. Although each case sounds preposterous to the layman, I think that abiogenesis is the least preposterous."

Back at the hotel, Blaise and Brian retreated to their rooms. The sessions started rather early in the morning, so they planned to meet for breakfast in the hotel coffee shop at 7:30 AM. Since Blaise was filled with food and feeling the late hour, he dropped off to sleep almost immediately.

10 DISCOVERY

Research on the mystery signal was at a standstill while Blaise and Brian were away at the conference, but they returned eager to determine the signal source. As usual, their first day back involved plodding through their e-mails, fishing through their travel bags to find receipts, and completing their travel reports. Finally, they were able to program another observing session, and Brian briefed the technicians on what needed to be done. The signal wouldn't be above the horizon for several hours, and this convenient lull occurred right at dinner time.

Blaise organized a dinner expedition with Brian and three others to a Mexican restaurant about twenty miles from the site. Normally, they didn't travel so far for dinner, but Blaise had heard good reviews from his colleagues at the local university, and he convinced everyone that it was worth the trip. Since his reputation as a bachelor connoisseur was on the line, he was as nervous as anyone as they approached The Lunatic Bandito. Happily, the parking lot was almost full, which was always a good sign. The restaurant sign, an image of a pistol laid across a crescent moon, looked much like the Soviet hammer and sickle.

There were five in their group, and since the booths were designed for four, there was a wait before they were seated. Someone joked that Brian, the low man on the totem pole, should just eat at the bar, so the rest could be seated immediately. The wait was just a minute. The waiters had pushed two smaller tables together, rather than waiting for a larger party to finish. From the looks of the crowd, the typical ratio of liquid to solid refreshment was about one to one, so not many planned to leave that soon. Blaise started to worry that the food reviews he heard may have been clouded by an alcoholic haze.

Blaise volunteered to be the designated driver, principally because no one was willing to take his ancient SUV out onto the open highway. In his younger days, he had bought it for off-road stargazing with the opposite sex, but his investment never paid off. Now he was stuck with a 200,000-mile-plus gas guzzler. If he didn't live so close to the site, he would have traded it in long ago. Now it was like an old friend. After the first round of drinks was served, the conversation got around to Brian's mystery signal. Tom, a veteran pulsar man, started taunting Blaise.

"I hear you and Brian detected some *I Love Lucy* reruns from a satellite somewhere."

"Yeah, it could be a spurious signal. The signal's so weak, there are a lot of candidates. It's possibly some intermodulation distortion in some satellite's subcarrier amplifier. Then again, you never know, that's why we've continued observations. One interesting signal we detected years ago turned out to be from a power line insulator, if you can believe that. The power line acted as an antenna that picked-up a signal from a leaky cable television line. Some acid rain pollution had built up on the surface of the insulator, and that turned it into a semiconductor rectifier. This rectifier had enough non-linearity to mix a couple of strong television signals into another frequency that was re-radiated from the power line. It took me a few weeks with a radio and directional antenna in my SUV to track it down. That was the easy part. Getting the cable TV company and the power company to fix things - that was the hard part. Of course, the FCC was no help at all. The signal was too weak for them to bother with. Of course, the fact that it was on our frequency, and not a *useful* frequency, probably didn't help."

"Don't you guys get tired of this, chasing down one false lead after another? I don't find a new pulsar every month, but I'm fairly certain that there are more pulsars out there for me to find."

"It's a matter of faith," said Brian. "Sure, we could do cookbook astronomy, like you guys..." Everyone laughed. "... but I need a little mystery to get me out of bed in the morning. That's what I think science really is, solving mysteries."

The food arrived quickly, another advantage of eating Mexican, and everyone chowed-down. Blaise especially enjoyed the lime flavoring of his beef burrito, and he decided that the restaurant reviews he heard were right. It was nearly 10:30 PM when they arrived back at the site. When Brian stepped from the SUV and walked towards the building instead of his car, Blaise called out, "Where are you going? You're about eight hours early for work!"

"I guess everything I said at the restaurant about solving mysteries is really true. I want to take a first cut of the data. Want to come along?"

"Sorry, Brian. Old guys like me need their sleep. That's why we hire post-docs! I suspect that the universe will still be there in the morning. I'll see you then."

One thing about a good Mexican meal, or a good meal in nearly any other cuisine, is that it has a soporific effect. Blaise needed to be attentive to his driving, and he was glad that there were very few cars on the road at that hour and that it was just a short jaunt back home. Some loud music on the radio helped to keep him awake, as well.

Although night driving had the advantage that there were fewer cars on the road, Tukey still hated driving at night. In fact, he hated driving, in general, and he was hoping that the self-driving cars he read about would come soon. However, on his salary, he would need to wait until that point, then a little farther until the prices had come down.

Still, there would be the problem that he would likely need to shut down the car's computer and drive in manual mode for the last half mile to the telescope. Perhaps the observatory would eventually build a remote parking lot with some sort of shuttle service.

Tukey had been in bed for several hours when the telephone rang.

"Blaise, this is Brian."

"Brian, I can't see the clock without my glasses, but I'm guessing it's not a proper time to call!"

"Sorry, but this is important. I've been looking at the latest data, and there's something here you should see.

Come in extra early in the morning. I'll be here when you get in."

Tukey forgot what his own work habits were when he was a younger man. "You're going to work through the night?"

"Yeah. It's almost morning anyways. The coffee isn't bad when you wash the pot first."

"OK. I'll be there before 7:00 AM. The guard will likely want to see some extra ID, since he knows it wouldn't be me at that hour!"

Tukey hung up the phone, rolled over and quickly went back to sleep.

11 FREQUENCY MODULATION

Tukey somehow managed to awaken an hour early. After a quick shower, he skipped his usual breakfast at home and jumped right into his SUV. As a concession to his nervous system, he stopped at a coffee shop along the way. He was surprised to see the shop open, but the number of cars in the parking lot was proof that there were quite a few people in need of coffee at such an early hour. He ordered an extra large Cafe Mocha, Tutto-Grande Size, or whatever they decided to call it, and a cinnamon bun that should have had a Surgeon General's warning attached about calorie and fat content. Blaise munched the bun during the two additional miles to the site, but he decided the coffee was too hot to handle in the car. Within minutes the familiar sight of the dish antenna greeted him.

The guard house at the site was manned twenty-four hours a day, seven days a week. Everyone here kept astronomer's hours. The gate house was there just for visitor information during normal hours, since a guard at the front door handled security from 7:00 AM through 7:00 PM. Beyond those hours, the entry gate was down, and this tiny house was the site's first line of defense. This morning a young Hispanic woman greeted him at the gate. She was too new to her job to recognize Blaise, or anyone else for that matter, so she carefully scrutinized his ID, and even checked the back of his SUV. Although he clearly wasn't a visitor, she gave him a copy of the Guidelines for Visitors card with information about not using cellphones, et cetera. He figured her new employee zeal would last about a week. She passed him through, and he used his ID card to buzz himself through the front door.

The front desk guard was just setting up for the new day, but he paused to greet Blaise.

"Doctor Tukey, your young man, Brian, didn't check-out last night. He isn't using the cable way as an alternate exit, like the last time, is he? It could be dangerous."

"No, he actually worked through the night. You know the type – no girlfriend and married to his work."

"OK. Just a friendly warning. We need to keep a tight ship. It would be good if he walked down here and left us a note when he was working late. I know you guys are dedicated to your work, never watching the clock, and all that, but I've got responsibilities."

"I'll pass along the message. I'm sure he will see the necessity. He's a logical sort."

The guard buzzed Tukey through the inner magnetic door. Brian was not in his office, but he found him in the control room. Surprisingly, Brian was hunched over an oscilloscope with one of the technicians.

"So, what's up, Brian?"

"Jeff here has been really helpful. He says things tend to be boring on the overnight shift, but my little questions have gotten his blood flowing again."

Jeff, one of the younger technicians, smiled and went back to adjusting some settings on the oscilloscope.

"Here's what we've been doing. Of course, I'm talking about the strange signal we've been studying. Your suggestion about tightening up on the bandwidth was right on the money. When we did it, things started to happen, but we didn't understand what was happening. Some of the channels started jumping around, like there was a lot of noise, and all the other channels were quiet. Finally Jeff suggested a reason. We're looking at a frequency modulated signal."

"Nature usually likes amplitude modulation. You mean Doppler from some orbiting object?"

"Too fast for that, unless we've found something really unique. But there's more. It's FM, but more specifically, frequency-shift keyed FM. I think it's an intelligent signal."

"OK. Time for a little Okham's Razor. If it's an intelligent signal, and, yes, FSK would qualify as an intelligent signal, then it's from good old Mother Earth. We just need to track where this interference is from."

"Exactly, that's what we've been doing all night. First, we examined the modulation. Obviously, we don't have an FM detector - no need for one until now - but we used the channelizer and some computer code to synthesize an FM receiver – you know what I mean. Then we got the modulation rate, 473 hertz. Sound familiar?"

"No."

"Maybe if I converted it to a period for you - 1.74 milliseconds.

"That's one of the millisecond pulsars."

"In fact, it's the closest pulsar in the direction of our signal source."

"So, we don't have man-made interference. Somehow the pulsar is getting into one of the sidelobes?"

"I don't think so. First, we're getting FM, not AM, and not just any FM, but FSK. I asked Jeff about it, and he said that some non-linear mixing somewhere in the receiver could possibly do that, so we're doing some upstream checks." Brian motioned towards the oscilloscope. "At this point, we've just about ruled that out. And while we were doing these checks, we were collecting the demodulated signal. It's a digital steam, alright. Not just a continuous square wave, but long pulses and short pulses, looking like typical digital data. I think that by locking onto the pulsar frequency, they're making it easy for us to detect."

Tukey was starting to get flushed. They! He could hardly believe what was being said. His inner scientist was searching through all the possible devil's advocate scenarios, but Brian had skillfully anticipated all his arguments.

"We're doing one last check. I asked Stanford to record ten minutes of the pulsar itself at the same time we record. It's a relatively strong signal, so there was no problem for them to acquire it with their small telescope. Their digital converters are down, and they couldn't send me the data over the wire, so I sent Marianne to pick up the tape. This is just a double-check. I'm sure it isn't an artifact. The idea that the pulsar would skip pulses is pretty far out, but it's an important check. Hey, if it isn't a SETI signal, missing pulses on a pulsar is a big enough discovery!"

Marianne retrieved the Stanford tapes, and there were no missing pulses. The signal they had discovered was locked to the pulsar's signal, but it wasn't just an echo of the pulsar's signal. The rest of the morning saw frenetic activity from Brian and Tukey, who at this point kept the potential discovery to themselves. Brian, who hadn't slept, was running on the adrenaline.

The first order of business was to record as much of the signal as they could. In theory, the signal could be received at their telescope for about five hours each day, but only if Brian and Tukey had complete control of the telescope for those hours. Because of the nature of their work, they typically operated at those odd hours that no one else had scheduled, choosing whatever targets of opportunity were in the sky at the moment. Now, they would need to be on the schedule.

The first problem with this is that they really didn't want to reveal why they needed to be on the schedule. This was news that should remain secret until it was confirmed. The confirmation was two-fold. Brian and Tukey needed to receive enough of the signal to convince themselves that it was true, then they needed to get another telescope on board to independently confirm the signal.

The next problem was that others had already scheduled for time at the next, and subsequent, periods when the mystery signal could be received. Tukey had some seniority, and, more importantly, funding. However, he didn't want to strong arm his way onto the schedule. That's not a good way to do science. He would use diplomacy. As a last resort, he would appeal to the director. He hoped to avoid this, since the director would want an explanation.

His next observation window had been scheduled by Jens, a rotund and affable character more likely to be seen in the coffee room than at his desk. Jens would be an easy target, and Tukey found him, as expected, in the coffee room. First, the small talk.

"Hey, Jens. Your countrymen made a good showing on the football field last weekend!"

Tukey had carefully called the game football, and not soccer.

"Yes, the team won, but that part about 'countrymen' is all wrong. Most of them are ringers, and they likely wouldn't have been allowed entry to the country if they didn't have golden legs. They're just paid help. But, Blaise, enough about football, hear from any green men, lately?"

Jens laughed, but for once his jest might be without merit.

"Perhaps, someday... But I've come here asking for a favor. My student has a conference abstract due, and he needs to tie off just one small loose end. Can we use your scheduled time tomorrow?"

"Yes, but you must reciprocate by bringing me a real coffee tomorrow morning...." Jens motioned towards the coffee makers at the back of the room. "...Not like what those chemical factories spew! Make it a caffè latte."

"Fine. That will give me an excuse to stop at the coffee shop and indulge myself, also. And a sticky bun?"

"You Americans and your thousand calorie breakfast treats. Bring me a croissant, a little butter on the side."

"You've got it," said Tukey, and he rushed to tell Brian the news. He advised Brian to get some sleep before then.

Tukey needed some sleep, himself, but he was obligated to handle a few office tasks before leaving the observatory.

The first task was to schedule an alarm to remind him to get Jens' coffee and croissant the next morning. Tukey was a man of his word. While others could just enter the alarm into their smartphone, he couldn't use his phone at work. Tukey needed to do a workaround in which a program on his office computer signaled a device that would update his phone wirelessly when he arrived home.

Tukey also needed to respond to a few email messages from colleagues. One of these, a request for some data, would need to wait for another day when he had more time. As a courtesy, he responded that he had received the data request, but it would take a day or two. As he composed the message, Tukey knew that verification of the signal would prevent his fulfilling this request, but his colleague would definitely understand why.

That afternoon, Tukey traveled home on autopilot, his mind consumed with the possibilities brought about by this SETI candidate signal. He outlined a schedule of how events might unfold. Tukey tried to live an uncomplicated life. Now, his routine was shattered, but he didn't mind.

As he entered his house, he immediately checked his cellphone to make sure his alarm had updated. He was more concerned about getting Jens his coffee and croissant than what he would need to do when his life's goal was thrust so unexpectedly upon him.

12 PROTOCOL

"OK," said Brian, "I've printed the SETI Protocols from the Internet. Would you believe that our library didn't have a copy?"

There were three of them in the main conference room, a room usually reserved for larger meetings, but it was needed in this case because of the video conference facilities.

"I can believe that," said Kris, "They were only of academic interest until a few days ago."

Kris Singh was the institute director. He had gained prominence in radio astronomy by championing, and finally building, a world class instrument in his home country, India. The Large Area Pinpoint Array, as it was called, was much like the telescope Jocelyn Bell had used for her pulsar discovery, but it covered a much larger area. Operating on a frequency ten times that of Bell's instrument, it had super resolution, but it was at the mercy of the Earth's rotation to sweep the beam across the sky. Located at Bangalore, near the equator, it was able to view most of the galactic plane, but it was nearly impossible to select a specific object of study. It was a survey instrument.

Politics in India had always been an impediment to its science. Some countries, like India and Italy, had never gotten past the tribal mentality that seemed to be inherent in humans. A major scientific enterprise, like the Large Area Pinpoint Array, required a lot of backstage wheeling and dealing. Kris had played the proper games when he was younger, but it had started to wear him down. His repeated attempts to modernize his telescope to allow it to track objects over a range of angles using computer techniques were rebuffed. Kris' ambition eventually transcended the resources of his still developing country, and he had accepted his position at the institute when it was offered.

Kris continued, "Now that we're here, let me scratch out an agenda. George Berkowitz of AURA, our putative boss, will join us via video in a few minutes. I telephoned George very early this morning, just an hour ago our time, but a bit later by George's time. He was still in transit to his office."

Tukey interjected, "Quite appropriate for an astronomer to be 'in transit...' " Everyone laughed. "...But what did you tell him?"

"I didn't get into specifics. You have to assume that open cell calls are not secure, although I don't know why anyone would be interested in eavesdropping on George, or me, for that matter. I told him we made a significant discovery, and we should discuss press release protocols over the video link."

The video link was secure. It was not as if it was ordered that way because astronomers needed secure video conferencing. The telecommunications company that designed it sold it to all sorts of businesses, and the security was built-in by necessity of the marketplace. There was no need for astronomers to keep secrets. The more publicity they got, the better the funding situation. It's what kept NASA in business for so many years.

Kris continued, "As for the agenda, we'll have the usual introductions. I'll spring the 'Big News,' field the expected questions from George about checking the data, etc., and then I'll have Blaise present the Protocols. We need to impress on George that we can't make any announcements until some independent group confirms the signal. Any suggestions whom we should call about this?"

Tukey answered, "I think Effelsberg is an obvious choice, being outside the country, and all that. We've done work with Klaus over the years on very long baseline interferometry, and I know he'll confirm the signal in less than a day. His facility is quite disciplined... well, maybe I'm relying on some German stereotype here... but if we choose Klaus, I'll need to contact him immediately after the meeting. The signal will transit his longitude in just a few hours."

"OK. It'll be Effelsberg."

Just then, the video switched from its standby screen to an image of George Berkowitz.

"George, we're receiving you! How's the weather?"

"This is Boston. Do you need to ask?"

"Don't feel bad. We must have had at least ten minutes of rain here, this morning, and the temperature will only peak in the seventies!"

George didn't appear to be a morning person, and he had just fought Boston traffic for half an hour. He wasn't amused. He sipped at his coffee.

"So, what's your news?"

As Kris eased into the discovery, George's face changed in incredible ways. Kris' monologue awakened George much more than coffee could ever do.

Kris concluded, "We've done all the necessary checks, and it looks like the signal is genuine. I've suspended all other activities in the five hour window the signal can be received each day, so we can record as much as we can. Now it's time to get others involved."

Gorge's mind was racing, planning, and scheduling. "Kris, this is big, really big. I've got to call a press conference. It will be at your site. I'll get a flight immediately."

"Having you at our site would be great, but you'll need to wait at least another day for the press conference. Have you heard of the SETI Protocols?"

"Kris, I'm just an administrator. I stopped doing astronomy years ago, and then I was an optical astronomer. I don't know much about what you radio guys do, let alone the smaller SETI community. What are these protocols?"

"I'll defer to Blaise on this one. Blaise..."

"Well, something like this was anticipated, or maybe I should say hoped for. At the end of the last century, astronomers agreed on a fair and logical way to handle this. It's a list of guidelines called the SETI Protocols. We'll e-mail you a copy. It's only a couple of pages. We've already done the early stuff, checking to make sure we haven't fooled ourselves. The next step is to get independent confirmation. We're going to contact friends at Max Plank in Germany to do the checks. If all goes well, this will take a little less than a day. But... and this is a big but... until then we don't tell anyone. When we announce this, it needs to be real."

George's face dropped. "OK, I'll agree with you there. Imagine the other kind of publicity we would get if you were wrong! When things are arranged, I'll fly there for the press conference. Kris, call me immediately on my cell phone when word comes back from Germany."

After a few parting pleasantries, Kris shut down the video link.

"OK, our path is clear. Blaise, contact Klaus immediately. Blaise and Brian, get together some data for the press conference. I'll have our artists put together a presentation. Then there's the press release... Well, I can put that together from the presentation. I think I'd better give what's-his-name from The New York Times a heads-up to travel here. I guess he'll have to take my word it's important, since I can't tell him anything. It's likely he'll get a leak from Germany... we can't plug all the holes... and he'll arrive forewarned and forearmed. Actually, I'm surprised he hasn't called me already!"

Everyone laughed again, but Kris' message was clear. News like this can't stay secret for long. As they were leaving, Kris called out to Blaise.

"Blaise, there's one other thing we need to discuss before you contact Klaus."

Brian filed past Blaise, and the door was closed.

"Blaise, there's one thing about the presentation that we need to discuss. If I didn't bring this up, then George would, so I think it's better we discuss it now, as colleagues."

"Kris, do you mean you're going to ship me off to Tasmania if Klaus can't confirm the discovery?"

"Hardly. There has been a discovery, one of the greatest achievements of mankind, and I have no doubts about it. I also have no doubts about this, that *you* are the discoverer."

Blaise was silent for a few seconds, collecting his thoughts. His final reaction, however, was more instinct than reasoned argument.

"Bullshit! What about Brian? If it wasn't for his work, there would have been no discovery! Brian can't be left out of this!"

"Blaise, we're not going to bury Brian. His name will be next to yours on the presentation, and he'll be at the front desk, likely between George and me, and maybe the Governor, for that matter. You and he will coauthor all the major technical papers. But for the purposes of the announcement, you're the discoverer. You're the project leader, the ultimate decisions were yours, and we need to speak from a single focal point, and you, Professor, are it."

Blaise thought about some rebuttal, but instead remained silent. Kris was right. It had happened to Jocelyn Bell, and now it was happening to Brian. Somehow, this was the way things were done, and somehow Brian would understand. At this point, the only thing Blaise could do was to mention Brian as often as possible.

Kris continued, "I'll have a talk with Brian about this, but not just yet. There's too much work to do before the press conference, and we can't afford to mix emotions with science."

Blaise left the conference room and retreated to his office. He needed to talk to Klaus at Effelsberg, but not just yet. He wasn't a drinking man. Even if he were, alcoholic beverages were banned from the laboratory. He grabbed a candy bar from his bottom desk drawer and drowned his anger, and a little guilt, in chocolate. These LGM were getting to be a lot of trouble.

Klaus was not in his office. It was after hours, German time, but Klaus kept his own hours, like Tukey. After some maneuvering through the switchboard, Tukey was able to connect with one of Klaus' students. Fortunately, the student remembered Tukey's name, probably from that paper he had written years ago about meta analysis of channelized signals that was now considered a classic in the field. There were so few papers in the early years of his field that every second paper became a classic. The student gave him Klaus' cell number.

"Klaus Mueller, Guten Abend."

"Klaus, this is Blaise Tukey. Good evening to you."

"Blaise! Such a surprise! Sorry I missed attending the last conference, budget problems, and all that. We could only send a few people. Definitely next time. To what do I owe the pleasure of your call."

Klaus, always the proper gentleman.

"Klaus, I have some coordinates for you."

"Blaise, I'm in my vehicle, going for an early supper. There's a nice little swiss restaurant a few kilometers from the site. I'll take you there when you visit again. Have you caught another pulsar? I'll be in my office in about an hour. Would you call then, when I have a pencil in hand?"

"Klaus, this is a little more urgent. Is there a place for you to pull over?"

"Sure, there's a little shopping center parking lot up ahead. Reception should be good because they built a cell tower right in back of it. Of course, we tried to stop them, but how can you stop progress. Hang on."

After a short pause, Klaus' voice returned.

"OK, Blaise, I have a pencil and paper. What's the emergency?"

"First the coordinates – 02 hours 30 minutes, Right Ascension; +42 degrees, 30 minutes, declination."

"That would place it in Taurus?"

"Close, it's in Velpuncula."

"Hmmm, it should be above the horizon shortly. Clear skies tonight. Almost makes me wish I were a *real* astronomer."

"Klaus, you need to scan it tonight."

"Tonight! I'm not on schedule tonight. It can't be done."

"Klaus, you need to pull some strings for this." Tukey paused, though not for emphasis. He wondered how exactly to phrase it. "We have a potential SETI candidate."

Now it was Klaus' turn to pause. "Blaise, to get a call from you, at this hour, on my cell phone, convinces me that you think it's more than just potential. What are the details?"

"Well, there's a lot of important stuff you need to know about the signal, but that can wait until you arrive back at your office. This won't be a routine measurement. You've got to change some electronics. There's some frequency modulation."

"There's some Doppler?"

Blaise paused for a second, knowing how Klaus' would react. Good scientists are born skeptics, and Blaise had found that German scientists excel in both their science and their skepticism. "It looks like data modulation."

"FM data? It must be an artifact, Blaise. Some satellite... something like that. How could we be so lucky?"

"Do you have anyone on your technical staff who speaks English well?"

"Better than that, we have a real Englishman. He's on loan from Jodrell Bank for some experiment. But most of us know English quite well, or at least as good as our French!"

Blaise took the jab in stride. Americans are notoriously monophonic, while educated Europeans are multilingual.

"What you need to do now is twist some arms, get on the scope for at least a couple of hours, and check this out. Your observational window starts in a few hours, so we need to work fast. Can we get a teleconference arranged in about an hour, your people and mine?

"Better make that an hour and a half. I'll need to find everyone. Some may have already left for the day. Let's see, that would put it at 6:30 my time, or 11:30 your time."

"Correct. I don't need to tell you that we're keeping this private for now. As you said, it might be an artifact, some spurious signal. The fewer on your end who know, the better. If you need to contact your director, OK. But it would be better if you just made the arrangements with whomever you need to steal the time. You could tell him you're under an abstract submission deadline, or something like that. I don't need to remind you that Max Plank, and you personally, will get significant publicity from this if we're correct."

"Blaise, a typical experimentalist, understating your case! If you were a theorist, you would speak with more certainty! We'll be drinking champagne from paper cups before daybreak!"

"... or we'll be making apologies all the way around. Thanks for the vote of confidence. Talk with you in an hour and a half. Call me at my office. We'll be expecting your call."

Tukey called Brian, and Brian said he would talk with Jeff the technician. Tukey decided to limit the conference to just these essential three. He and two others would just fit in his office, and he would rather talk from there, with the door closed, than in a conference room.

At the appointed time, Tukey's computer beeped, signifying some significant event, which, in this case, was Klaus calling on videoconference from Effelsberg. Tukey checked for the presence of encryption on the link. Assured that it would be difficult for anyone to eavesdrop on the conversation, he clicked the button to answer. He saw Klaus and a few others in a small conference room. The door to the room was open.

"Hello, Klaus! Could I first ask that you close the door to your room?"

At this request, the conference team member closest to the door immediately rose from his chair and closed the door. Klaus responded.

"Sorry, Blaise, we usually keep doors open here as a compromise to the faulty air conditioning. All power goes to the 'scope, so to speak!"

"Similar funding story here, I'll admit," said Tukey, then he introduced Brian and Jeff. Klaus likewise made introductions on his end.

"Klaus, since Brian and Jeff have the details, I'll let them summarize. If you have any questions, interrupt at any time. After this meeting, send me an email with everyone's contact information, and I'll reply with the same information on our end. I'll also include the frequency and source coordinates, so there won't be any confusion. Brian, would you start?"

Brian, working from his notes, along with some embellishment from Jeff, described how they had been able to receive the frequency modulated signal using a tight bandpass filter and what Jeff called "slope modulation." The Effelsberg team immediately understood the approach, and after a few minutes of discussing, sometimes in German, how the same could be done at their telescope. After a time, Klaus gave a status report.

"Yes, it can be done, here, and quite easily. It's just a software command. Ah, the miracle of computers... No screwdrivers required. Not like the old days, right, Blaise?"

"Klaus, sometimes I miss handling the hardware; but, who can argue with progress, at least not in our profession."

The teleconference ended, and now it was a waiting game.

It was no use trying to sleep. Klaus would call Tukey on his cellphone, so it didn't matter where he was at the time. As a courtesy to Brian, Tukey went out for a pizza after the teleconference, brought it back to the observatory, and remained in his office for the duration. Tukey was quite certain that the FM signal was no artifact, but reception of the same signal half a world away on a different telescope was very nearly definitive proof that it was a SETI signal.

Tukey's cellphone rang. He glanced at the wall clock as he answered. It was a little after 1:00 AM, so the Effelsberg team had a an opportunity for more than an hour's reception at that point. The Sun would have just risen at Effelsberg, but radio astronomers didn't keep solar time. Any hour of the day was useful time, provided that their target was comfortably above the horizon.

"Klaus?"

"Blaise, we got it! No problems at all... Such a strong signal. We need to compare data."

"Great! I'll tell Brian, but all of us deserve a rest at this point. After a six hour break, I'll give you an uplink address where you can pull the raw data. It's only a few Gigs."

"Thanks. I'll do the same. Blaise, you'll be famous!"

"Famous, maybe. Tired, yes. I'll consider my fame after a little sleep. Remember, we're to keep this quiet for now – SETI protocol, and all that. Advise your team in no uncertain terms not to leak this, but you'll need to tell your director so you can get this observation time slot every day. It'll be known soon enough, likely in a few days. Then there's the little matter of comparing our data, which hasn't happened yet. We need to be thorough, so we don't end up as laughing stock. Above all, remind everyone about the SETI protocols."

While walking to Brian's desk to announce the news and plan for the data sharing, Tukey slowly began to realize that this was just the beginning of a long saga. He'll get very little sleep for the next few weeks, perhaps months.

Tukey was surprisingly rested after just four hours of sleep. That was all he had allowed for himself. Not long after awakening, he arrived at the observatory, not expecting Brian to be there. Brian, however, was there, and from his appearance, he had just slept in a chair overnight. He was working on getting the SETI data uploaded for Effelsberg's access.

"Professor Tukey, I'll have this ready in just a few minutes. I'm sure the Effelsberg people will be just as prompt, and I'm eager to put both data streams on the display for comparison."

"A cross-correlation would be good. It's good to check that, but it's unlikely the same data is in both data streams unless it's just a short, repeated message. More certain would be the pulse rate similarity, and how it matches to the pulsar. I'll be in my office. Email the link, and I'll take it from there."

When Tukey had Brian's link, he added it to the email message that he had already composed for Klaus and his team. Almost immediately, he got their message in response. Klaus must have a "Brian" on his team. Jeff was copied on the message, so Jeff now had the link to their data. Nothing to do in the interim except wait for Brian's call.

Although Tukey tried to go about his usual activities, his efforts were only half-hearted. It wasn't from lack of sleep. It was from Klaus' remark about his pending fame. He was an academic who had fallen into a comfortable lifestyle. He was well paid, at least as far as his profession allowed, he had an interesting work experience, and most mornings he was happy to arrive at work. If, as Klaus had said, he would be thrust into the limelight, all that would be changed. Would he be happier? He doubted that, although he would smile in his later years when finding his name in some magazine article or history book.

Brian was as efficient as always, and Tukey heard a knock on his office door after just a half hour.

"The signals match - Well, at least the frequencies match to the level of statistical error you would expect for the number of bits in the smallest data stream. There's no correlation, meaning the bits aren't the same message, so these samples are likely pieces of a longer message."

Now that he heard the word spoken, Tukey realized that not only had they discovered the existence of an extraterrestrial civilization, but they were getting a message from them.

Brian continued.

"So, what's next?"

"Next," Tukey responded, "is somewhat out of our hands. Kris has granted us the observation slot tonight, and every night, and I'm sure Klaus is working the same at his end, discretely. We need to get other observatories to fill in the gaps when the signal source is below our horizons. Kris needs to start planning to make this public. Maybe Klaus and I can enlist the Australians as another covert receiver, but this can't be done properly without making it public."

"And that means the SETI protocols."

"Right."

13 DINNER

Friday dinners alternated between Rick and Tom's, with the wives handling the cooking. This week, dinner was at Rick's, and Mary was sure to come through with an excellent, and often exotic, meal. The last time, Mary had cooked Indian food, chicken vindaloo with a kheer dessert, a good combination of hot spice and cold. These dinners were a way to unwind after a long week, but in most cases the week was more likely to run from Monday through Saturday, rather than Monday through Friday. With the main course on the range top, Rick, Tom, Mary, and Constance, had a few moments to relax over drinks.

Tom's wife, Constance, was usually the life of the party, and she started the conversation with a joke she found on an Internet newsgroup.

"There was this MBA type," she said, looking across at her husband, the apparent butt of this joke, "who was doing a project to increase farm productivity. He visited a farm, and saw the farmer out in a field tending to his sheep. After a simple hello, he proposed a wager to the farmer. 'If I can tell you how many acres you have to support a flock of this size, would you give me that sheep?' The farmer said yes. 'Twelve acres.' 'That's exactly right,' said the farmer. The MBA took his sheep."

"'Will you give me my animal back if I can guess who you are?' said the farmer. The MBA agreed. 'You're an MBA with no farm experience.' 'Right,' said the MBA, 'but how did you know?' The farmer replied, 'First, you came here like you owned the place. Then, you demanded a high fee for a simple task. Finally, you told me something I already knew. Now, give me back my dog!'"

There were laughs all around.

"But, remember," said Tom, "It's all those MBA skills that pay the bills."

"Yeah," replied Mary, "And it's Rick's shepherd skills that keep you focused."

At that point, the timer beeped, signaling dinner, so three of the foursome headed to the dining room while Mary headed to the kitchen.

Mary had learned to cook from her mother. Her mother's eclectic tastes resulted in family meals outsourced from nearly every world culture. There was fried tofu and various seafood specialties from Asia, a lot of Italian, Southern Indian entrées, Mexican, and even corned beef and cabbage. Tonight, Mary had made lasagna with a mix of sweet and spicy sausage and a generous proportion of several cheeses. The group liked this entrée at a previous dinner, and they were sure to like it again.

When the meal was over, Mary brought out a special treat that she had purchased at the local Italian bakery, a cannoli cake, which was essentially a cake glued together with cannoli filling. Those who insisted on a small piece at first, asked for seconds. Mary's dinner menu was a success once again.

Even in these days of women asserting their rights to be more than just wives and mothers, the women still retreated to the kitchen after dinner, while Tom and Rick headed to the living room. With drinks in hand, they plotted the near term course of their company. Rick started the conversation.

"I'm happy we hired Dan. Maybe we would have burned through all our money at this point if it wasn't for him."

"Me, too, but we didn't hire Dan, remember? He was forced onto us by the money guys, but I'm not complaining. We're getting close to going live, and we have enough money in the bank to see us through until then."

"Then what?"

"It's the usual story. Second round financing after we show some results. Then, much farther down the line, an IPO. After that, we alternate dinners between my yacht and yours, but I'll insist that Mary cooks and not her staff."

"Yeah, the American dream, but I enjoy my work. Physicists aren't in it for the money. They want to do interesting things."

"Then you'll start a research foundation, or fund another startup. It's the usual serial entrepreneur cycle."

Rick thought for a moment, then he asked Tom, "You don't intend to just sit on your backside after this is over, do you? What are your plans?"

"Yeah, the television depiction of the idle rich doesn't sit well with me, either. Intellectually motivated people need to do something with their time. It's nice to know that you can do anything, anything at all, since the object is not making a living. I think I'll write."

"Books? I never thought you were the type, although your white papers and memos have a certain style."

"As you've noticed, the out-of-office message in my email is well crafted."

They laughed, and Tom continued.

"I would never write one of those "business methods" books. You know, *Here's how I made my millions, and how you can do the same.* Things change so rapidly in the world that yesterday's winning formula would never work again. I'd write novels, probably historical fiction. I've always enjoyed reading those, myself, so I think I would enjoy writing them. Those books have substance, since you need to do some research, not type page after page off the top of your head."

"I was sort of hoping you'd write romance novels. Then I could get you to put an artist's rendition of me and some babe on the cover."

Constance and Mary, wondering what all the laughing was about, entered the living room.

"You guys must have had too many drinks, already," said Mary. "It's time for Connie and me to play catch-up."

With that, the women snuggled with their husbands, and the conversation turned to more general matters.

"While you guys were talking shop," said Connie, "Mary and I were talking house. More specifically, playing house. We want to have babies."

If the men weren't attentive at that point, they suddenly became attentive as Connie continued.

"It seems like the time for excuses is over. You've got good jobs, we have nice houses, and Mary and I are home all day. Mostly bored." She grinned. "We thought we would give you the right of first refusal at being the fathers; but, after that, we look at other options."

Tom didn't know how to respond, so it was up to Rick.

"Other options?"

"I heard there's a Nobel Laureate sperm bank, or something like that. It might not still exist. But I remember that one of their Nobel Prize donors was a little wacky, so that might not be the best route. So, what do you say? If we coordinate this, maybe Mary and I can share a hospital room."

Tom was happy he still had a drink in hand. He took a long sip and responded.

"Well, I can't speak for Rick, but you have a valid argument. If not now, then when? A lot of our school mates had children while still in college. It was a challenge, but they survived. I don't know about coordinating conception, but we'll let nature take its course."

"I guess I'm in, too," said Rick. "When do we start?"

"As for Tom and me," said Connie, "As soon as we get home. Mary has the advantage that she's already home."

At that, the two girls snuggled closer to their men.

"There is one problem," said Mary. "In my case it would be a difficult delivery."

Connie seemed concerned.

"Why?"

"Our child would definitely be born with a computer in his hands."

Everyone laughed.

"Connie will have an easier time," said Tom. "Cellphones are much smaller than computers."

14 INTERNET BUZZ

Mary and Rick had met Blaise when Rick was still a graduate student. Blaise had arranged a year's sabbatical with a renowned radio astronomer at the university, but the astronomer died in a boating accident a few weeks before Blaise's scheduled arrival. Blaise arrived early to give a eulogy, and there he was, unmarried, out of his element, and very much alone. Rick attended the ceremonies, bringing Mary. Although Blaise was many year's her senior, Mary's motherly instincts kicked in, and Blaise became the newest addition to their circle of friends. Blaise remained for his sabbatical year, shepherding the now fatherless graduate students until another professor could be hired. The generation gap, and the fact that Blaise's research interests were far removed from Rick's, made him an interesting member of the social group. The fact that Blaise was professionally established, with a bank account to match, was an added attraction to the band of starving graduate students. The weekly pizza party evolved into an international food festival with the help of Blaise's generosity. They had kept in contact during the intervening years. E-mail had that blessed effect.

Mary didn't hear the news from Blaise Tukey himself. Blaise must have been far too busy these past few days to have slept, let alone electronically chitchat with his friends. Mary had received a forwarded e-mail from a friend who had received it from a member of Blaise's team. The message did not explicitly report detection of an alien civilization, but talked around the idea so skillfully that there was no mistake. It also stated that a press conference was planned. The time stamps in the message header indicated that the news was traveling fast. I would likely hit some message boards in the next few hours.

Quite likely, Blaise hadn't accessed his e-mail or chat accounts in a few days, so there was no use contacting him via those routes. In this age of efficient search engines, his e-mail address was easily known to everyone who might want to contact him, from the Vatican to the Lunatic Fringe. A private e-mail address must have been arranged for him quickly and communicated to essential contacts, but Mary wasn't on the "A"-list and wasn't notified. Mary took a chance that Blaise was still answering his cell phone, a far less public number than his e-mail address.

"Tukey here..."

"Blaise, this is Mary Christoffel. How are they hanging?"

"Mary! A fine way for a Lady to talk!"

"What do you mean, Lady? You know who I married!"

"Point well taken. So, how have you and Rick been?"

"Rick's essentially married to his start-up, now, and I'm playing the long-suffering wife. But enough about us. We heard the big news!"

"I'll be hosted at a press conference tomorrow. They'll claim that I made the discovery, Principal Investigator, and all that. It was actually my bright post-doc who made the physical discovery. I'll try my best to mention his name as often as possible. It might boost his job prospects, but I'm afraid all the glory will descend on me."

"Blaise, don't be so self effacing! You did your time, played the game, and now you reap what you've sown. Enjoy the moment!"

"Yeah, maybe you're right. I needed that. I want to talk more, I really do, but I'll have to cut you short. There are some people here, you know what I mean, so I'll catch up with you and Rick when I can."

Mary said good luck and good bye, and that was that.

The press conference came late the next afternoon. The timing was possibly dictated by market closing on the East Coast. The press conference was held at a local hotel, and there were possibly more cameras in attendance than people. There was no Internet video link, since the laboratory management realized that their servers would be so swamped as to be ineffective.

There was a late idea to host a video stream on a "cloud," but the IT people couldn't figure out how to do something like that live. A static video would be posted immediately after the press conference. The proverbial cat was definitely on the loose by the time the conference got underway, since the university servers were still swamped with hits from people looking for a video feed that wasn't there. It was "The Cyber Attack of the Little Green Men."

The press conference was booked for the largest hotel ballroom available. Even then, there were video links to several smaller rooms, with one huge monitor set under the awning of the hotel entrance. The Institute seemed to spare no expense for the occasion. Much to the consternation of the hotel guests, the parking lots were packed with equipment vans of the various television and Internet news services.

Possibly the entire police force was deployed around the hotel, but not just to direct traffic and keep order. There was also the imminent threat of a disturbance by the Lunatic Fringe. Fortunately, from the Internet message boards, the Fringe membership hadn't decided whether a SETI message was a good, or bad, thing.

There were arguments on both sides. On the one hand, there was the *evil alien* faction who believed that "we had better not answer, because it will give away our location." There was the *evil message* variation, in which the aliens seek to destroy us by planting dangerous information. There was a religious right version of this in which the message was a test from God, so we had better be careful.

The other side of the argument believed that the aliens were not just advertising their existence, they were also giving information about how we could survive the many crises that their civilization faced and overcame. The *good aliens* were interested in the welfare of their fellow sentients, so they devoted such a large measure of their resources to spread their good news.

Tukey, and most of his colleagues, just took the message as what it appeared to be, the calling card of another civilization. Some of his colleagues also believed that the radio message was just the least of a technological hierarchy of communications channels these aliens were using. They were using this quaint communications form just for the edification of their less advanced neighbors.

Demonstrators, if there were any, must have been pushed back by the crowds of newsmen, since Tukey saw none of them as their SUV, surrounded by police cruisers, delivered Brain and him to a side entrance of the hotel. The other dignitaries possibly had a secret service detail guarding them, which seemed incongruous to Tukey, since he believed that Brian and he were the most important speakers. Of course, no one knew what Brian and Tukey looked like. The Governor, however, was a different story.

They arrived about five minutes before the scheduled start of the press conference. Tukey was worried about the shortness of time, when one of his handlers, a press agent from the university, told him that press conferences rarely started on time, especially those attended by many traveling dignitaries.

Tukey and Brian were taken to the "green" room, which actually had a red motif. The carpet had a faux Persian design in red and blue, the wallpaper was a red and gold brocade, and the chairs had a red covering. Also in the room was their director, Kris Singh. George Berkowitz, Kris' boss, wasn't there. Tukey addressed Kris.

"Kris, George isn't here?"

"A short, long story. His flight was delayed, thunderstorms in the Northeast, so there wasn't even an alternative airline. We'll need to proceed without him."

"Who else will be speaking?"

"The governor is here. He'll be introduced by the university president, then it will be me, then you. We'll all be seated at the same, long table - very democratic, although I think we technical types will be on one side of the podium, and the politicos will be on the other. Makes it easier for the cameramen, or something like that. Are you ready for your fifteen minutes of fame? Oh, we have your presentation slides ready. Just don't be too technical. There's a press release to cover all that."

With that, there was a flurry of activity in the ballroom as the governor entered.

"Time to take our seats," said Kris. "Just follow me and look for your name on the place cards."

"Well, I guess that my quiet life of an academic has just ended," thought Tukey. "Let's hope that the benefits outweigh the cost."

The press conference went as they had expected. There were questions from the press about whether the space aliens were on their way for a visit, when they would arrive, and whether we've replied to the message. How did they know our language so that they could send us a message, what language did they use, what does the message say, and what do these space aliens sound like?

Tukey had anticipated all these questions. Although he was able to answer with a straight face, he could see that Brian was trying very hard not to giggle. For the occasion, Brian had prepared a sound bite. It was a portion of the message recoded in the dual-tone signals of early computer modems. This was the highlight of the press conference, and the press corp asked that it be repeated several times.

Kris fielded questions of a programmatic nature, such as what will happen next, and how soon. He said that the first priority was to enlist the aid of radio astronomers worldwide to capture the signal when it was high in their sky. Every bit lost could be a bit lost forever. Kris mentioned Effelsberg frequently, per Tukey's prodding, and stated that this must be an international effort. The US could not monopolize reception, since the signal was there for anyone to receive.

Just as it seemed that every possible question was answered, there would be another, then another. Kris kept things going for some time, but even he realized that his limit had been reached.

"Gentlemen, there will be a detailed press release with complete information available on our web site. New information will emerge in a matter of time; but, once things settle into a routine, hopefully within a week, our scientists...", Kris motioned to Blaise and Brian, "Will be available for interviews."

With the press conference over, Turkey and Brian were hustled out a back door and chauffeured back to the observatory. During the short trip there, Tukey was plotting the myriad things he needed to do next. There was a sense of urgency. Although his telescope and Effelsberg could capture the message for nearly twelve hours of the day, there was still a wide gap in message reception.

Kris, however, had pointed out that his data disclosure at the press conference allowed any observatory with appropriate equipment to start observation immediately, and that's exactly what they would do. He extracted a promise from Tukey that he would head directly home to rest for the remainder of the day, and not return to work until morning.

Ordered to rest and actually resting are two different things. Tukey, understandably wired by the events of the day and the discovery itself, wasn't able to rest, and he wasn't able to shift attention away from those events. He did his laundry, the dishes, and actually vacuumed the floor for the first time in many months. Finally, he settled down at the television, which was something he rarely did. He often thought that if his cable bill was prorated on an hourly basis that he would save a hundred dollars each month.

It was hard to escape his own images on the television. Some of the news presenters actually got a lot of the science right, although the expected counterpoints of potential space alien invasion were sprinkled among the channels. Tukey's estimation was that science had won the day, although the anti-science types were potentially dangerous. In any event, he had no fear that funding for message reception would ever be cut off, no matter which political party was in power. The message was just too important.

Fortunately, the observatory had taken the precaution of having his home telephone number forwarded to an answering service. Otherwise, it would ring off the wall from incessant calls from reporters and assorted lunatics. The people who really needed to contact him had his cellphone number, which was not public.

After a patchy night's sleep, Tukey arrived early at the observatory to handle some administrative matters impacted by the message. The message would consume all his waking moments, at least in the near term, so Tukey needed to offload his other assignments. This included finding a replacement lead for the Kumhansl Milliarcsecond Survey, and also finding someone to replace Brian as the working hands. Kris could just delegate these assignments, but Tukey didn't want to play the heavy hand. He decided to ask Jens.

Jens was an obvious choice. He always had difficulty getting funding for his preferred research topic, mapping interstellar gases through their microwave emission. The research was important enough, but it was relegated to the low end of the funding priority list.

Knowing Jens work habits, Tukey waited until 9 AM to look for him. Jens wasn't in the coffee room, but Tukey found him in his office. Jens immediately responded to the knock at his open door.

"Ah, the Man of The Hour; or, perhaps, The Century! It's good the discovery happened now, since the century is still young!"

"Thanks, Jens. Of course, Brian is the one who really discovered the signal."

"Blaise, I saw the news last night. If you hear it on television, then it's true! So, how am I so honored by your presence?"

"Jens, a small favor, but one that may work to your benefit, also. Would you like to take over Crummy?"

"Kumhansl? Sure. Not at all taxing, and it does pay the bills. By the way, has it been decided how your bills will be paid? The message is wonderful from a scientific standpoint, but the money needs to come from somewhere."

"Kris is looking into that. Right now, Brian and I are on overhead, but I'm sure the government would be happy to pour heaps of money onto this."

"Your same government that abandoned all SETI many decades ago? You should make them pay for that."

"Kris, and Berkowitz are good at politics. They'll cut us a good deal. Maybe we'll even get some equipment upgrades. Get your wish list in order."

"As the say in my country, *En stigende tidevand løfter alle både.*"

"Which means?"

"A rising tide lifts all boats."

15 FIFTEEN MINUTES OF FAME

Tukey had a life filled with television and other interviews for the first few months after the press conference while a hastily assembled team of radio astronomers scrambled to ensure that the message was recorded twenty-four hours a day. Tukey's involvement in the scientific aspects of *his* discovery (which he still referred to as Brian's discovery among his colleagues) was nearly zero during that time.

Tukey recalled the lamentations of more than one Nobel Prize winner that making an important discovery in science prevents you from doing science from that point forward. His eventual award of a Nobel Prize was expected, although the award lagged the discovery by many years. For controversial discoveries, such as the detection of the neutrino, the award was made decades after the experiment. Although the detection of the SETI signal was unambiguous, it was controversial in the sense that some people characterized the signal as evil. Tukey was embarrassed to think that he might receive a Nobel Prize for an accidental discovery, and a discovery made by someone else.

After his fifteen minutes of fame had blossomed into about two months, the SETI story retreated to the back pages of the newspapers. This was primarily because there was no further story. The signal was there, as confirmed through reels and reels of digital tape, but the message, whatever that might be, had not been decoded.

This hadn't deterred one artist who obtained a copy of a portion of the bitstream and rendered it in ones and zeros on a huge canvas. He had spaced groups of bits so that they appeared like words in a document, and he had colored the words differently. Although there was no logic behind any of this, the painting was a nice abstract image of the message, and it was picked up by the media.

The scientific work had been done by shifting funds from other efforts, mainly at the discretion of the directors of the various telescopes. However, everyone was agreed that a formal effort must be launched for both data collection, and for decryption of the message. Radio astronomers were good at decrypting technical manuals, which often looked like they had been written on other planets, but they were in no way trained to decrypt a SETI message beamed their way.

Not that they didn't try. The more computer literate converted programs normally used for signal analysis to look for patterns in the signal. The signal was quickly identified as intelligent, since the bit patterns were far from random. Not only that, but the patterns changed character from one section of the message to another, as if there were changes in topic, chapters in a cosmic message.

Radio astronomers realized that once they had the signal demodulated into its ones-and-zeros data stream, they were acting outside their area of expertise. The analysis of the digital signal was more like cryptography, a mathematician's game, than the purview of astronomers.

Cryptograms, the formal name for coded messages, were designed to elude decoding. The statistical nature of language, however, gave the cryptographer a small tool he could use to decipher coded messages. If they did use language as humans do, the senders of the SETI message wouldn't be so foolish as to send the message in their own language; that is, if the message wasn't one of their internal missives we had just stumbled upon. They would realize that such a message could not be decoded by an alien species.

The principal problem in decoding the message was the lack of apparent reference points, although there were conjectures as to what such reference points might be. The astronomers were certain that the local arrangement of stars with a pointer to the home world must be in there, somewhere. Other scientists were certain that the periodic table of elements would be there. Mathematicians were sure that mathematics was the universal language.

Some philosophers argued what appeared to scientists to be the completely outlandish point that the message might reference a science unknown to humans. What they meant by this was not some new discovery in physics, chemistry, mathematics, or biology, but another discipline that would join these on an equal footing. What might this be? Their point being that humans had no idea.

As Tukey had predicted, Kris starting working the funding agencies immediately after the press conference.

He had published a memo to his staff to develop a list of telescope upgrades. Although there should be some justification for these in terms of the SETI message, it was clear that everything short of a new dish was fair game.

What was becoming clear to everyone involved was that the message was stealing the thunder from many other research programs, much to the consternation of their principal investigators. After the initial flush of excitement, radio astronomers started to talk about calving message reception and its analysis from radio astronomy, lest it overwhelm the field.

Kris, working in concert with the directors of the prominent world radio observatories, devised a simple course of action. Kris asked Tukey to write an opinion piece for The New York Times suggesting that the appropriate path forward would be the creation of an international "Message Institute." Tukey's authorship assured publication, and he wasn't adverse to having his words edited into the florid prose expected in such articles.

The next problem was, what exactly to write. Tukey had written numerous papers for scientific journals, and after a while they had become very easy to write, since they were formalistic. You have some data, you explain how you got that data, you conjecture about what the data means, and you reference related publications on the topic. You never needed to turn a phrase, since it was all numbers connected by a few transitional words.

Since he had learned the art of writing scientific articles through the example of the many such articles that he had read, Tukey did the same for his opinion piece. He selected about a dozen opinion pieces on scientific and technical topics, and he read them through while outlining their construction. Not surprisingly, these, too, were formalistic, just adhering to a different structure. The important point was that an international effort was needed. The exact nature of the collaboration would be defined by more capable organizers than he. Kris approved the finished piece, and it was published almost immediately.

After publication of Tukey's article, prominent radio astronomers, at the behest of their directors, sent letters of agreement to the editors of the foremost scientific journals. The import of these letters was that this was a message to Earth, so it should be received by all the people of Earth. The "Message Institute" should be a true international collaboration.

16 SETI ENTERPRISE

Finally, many weeks after the discovery, and with the prodding of so much public opinion, the government decided to enter the fray, and team with the people that they had cut from their funding streams many years before. First, many stations of the extensive array of NASA tracking stations were enlisted to monitor and record the SETI signal. This helped to maintain data integrity when the signal was over ocean regions where just a single radio telescope had been recording the data.

And that was all it was, just data, as reels of magnetic tape, piled up at one location or another, with no responsible central authority to compile it into a single data stream for subsequent decryption. The urgency of doing this was underscored by recent history. The original data tapes that held video of man's first walk on the Moon had been lost, and only lower-quality versions were retained. No one wanted the SETI message to have a similar fate.

When it came to decoding messages, the US government always relied on its National Security Agency, which came knocking at Tukey's door for advice. Tukey's first advice was that the NSA, or any government agency in the business of keeping secrets, not just decoding them, should *not* be involved. The public outcry would be too great, even in the US, itself, let alone the world at large. Suspicions would run high.

Taking government facilities out of the mix created a problem. The government had access to many supercomputers, and supercomputers would be a requirement in any decoding effort. Although there were several large corporations capable of providing these services on a contract basis, these corporations had been accepting government money for decades; so, they too, would be suspect. Tukey had a suggestion.

That's why representatives from the NSA came knocking on Rick and Tom's door with a proposal. The government would pay for access to their computer system; and access, also, to Rick's computer expertise.

All this was easier said than done. "We have a business plan, and we've accepted money from investors," complained Tom.

"Sure, this is an adventure of a lifetime, and probably much more interesting than watching the stock ticker, but how could this be done?"

The deep pockets of Uncle Sam have a way of accomplishing things quickly, so the investors were bought out, at the hefty margin they demanded, and Tom and Rick became part of the SETI Enterprise. The first problem, of course, was what that part would entail.

In Rick's mind, the precise project definition wasn't that important, since every computer-based project had the same basic requirements. Memory was needed to store the data, which in this case was the data stream of the SETI message. Then there were user accounts by which multiple people could log onto the system to access the data. Finally, there were the application programs that these users would need to review and analyze the data.

Based on what Rick was told about the nature of the SETI message, he had more than enough memory. His problem was the users. His system was designed for just a few users, all of whom were located at the computer. What was needed, now, was a very high speed Internet connection, along with the necessary hardware to couple the high rate data stream into the computer. In this way, the SETI message could be uploaded from whatever radio telescope had acquired it for its designated reception time, and an authorized user could download the message from anywhere in the world.

Rick had never worried much about the infrequent power outages that would shut down his computer. In a cost-benefit analysis of their own arbitrage scheme, the money they would lose during these outages came nowhere near the cost of the expensive backup power system they would need. The SETI message requirements forced him to rethink this problem.

The message was too important for its integrity to be at the whim of a weather system or the local electric company. A hundred percent "up time" was needed; and, considering that they didn't have that much available real estate, a fuel cell system was the best option. The cost was considerable, but that's not what worried Rick. He was hoping that his government backers would be able to expedite the usually protracted permit process of their locality.

As Rick drafted his plans, he realized that the SETI message, although unique in itself, was fairly routine when it came to its computer requirements. Once things were operational, he could sit back and do some of the innovative programming that he enjoyed to assist in the decoding process.

Rick's operation was not technically a "front," but the government wasn't going to cede its interest in the message that easily. It would be quite easy to tap into the same data stream by which Rick received the SETI data, and then do its own analysis on its own computers, many of which were more powerful that Rick's. This would be a "black" program, well-funded but without anyone's knowing where the money was going.

It might turn out that the content of the message was just a mathematical encyclopedia, in which case the government would close its program and pretend that Rick's operation had always been the only game in town. They would feign amazement at the public release of a translated message that they had already decoded.

On the other hand, the message might contain information on a new energy source. That would be important in its own right; but, as the atomic bomb had shown, energy sources could be militarily useful. Then, the decision might be to corrupt the message to prevent decoding, although that might not be possible. Too many people had the capability to listen, there would be many copies, and the ruse would be discovered. However, even if the discovery was made, it could perhaps be blamed on some technical issue. In the end, a precious number of month's advantage might be gained in weapons development.

The SETI community of radio astronomers may have wholeheartedly backed Tukey's plan for a world Message Institute, but their governments weren't so sure that would happen. Some of them had pressured their astronomers to provide records of as much of the message data stream as they could obtain, and some governments were collaborating to piece together longer portions of the message.

As they were doing this, the cryptographers that they had enlisted for decoding protested that something like the message needed to be examined in its entirety, not as snippets. Snippets, they warned, would get you nowhere. One example that they used was a manual for defusing a bomb, for which the instruction, "clip the red wire," was read, but the rest of the sentence, "after first clipping the yellow wire," was not.

Such problems of attempting to independently decode the message finally reached the government policy makers, and the world governments had slowly become more receptive to a coordinated effort in a Message Institute. They realized that the United States, with its worldwide network of reception facilities, was clearly at an advantage. In forming a Message Institute, there would be little to lose, and much to gain.

Since the world was electronically interconnected, nearly a hundred percent of the population knew about the message, and they were personally considering what it might mean. The idea that there was "someone else out there," was disturbing to some, but reassuring to others. The fact that another civilization could age without extinguishing itself was significant to an Earth that was going through one crisis after another. If it wasn't another war, there were enough environmental issues to make people wonder if there would be any world left for their children and grandchildren.

Dogmatic religions transform at a glacial pace, so there were no tweets from the Pope about the message. Instead, he convened a council of scholars to establish the church position. The television evangelists weren't so reticent. Most of these said that it was another sign of the End Times.

Citizen reaction, the fodder of slow news days, was mixed. Everyone was interested in the contents of the message, and some were surprised that it needed to be decoded. After all, everyone speaks English in the Star Trek universe. There were some people who were genuinely upset about the idea of aliens interfering with Earth's culture. Tee shirts were even printed with a stylized space alien inside a circle with a line through it, and this had become the logo of a small clique of anti-message people.

There was even the idea that the message could be decoded by the public, itself, in a crowdsourcing approach. Since public expertise in SETI message decoding was questionable, a subset of academics might make a suitable "crowd." There were also proposals to re-purpose distributed computing schemes, such as SETI@home, for message decoding. After all, the SETI message was directed to all humanity; so, shouldn't all humanity be involved?

17 SUNDAY SUPPLEMENT

"You must be joking! I got a "C" in high school biology, and that's as far as I went in science. There must be someone better to cover the story!"

Amy Pomronger was again confronting her editor, Andrew. Conversations like this had gotten to be a habit, so he didn't seem to mind. Amy was one of his best journalists, and feisty was par for that course. He wouldn't admit it, but he rather enjoyed these verbal sparring matches. It reminded him of his early days in the press room when the writers, all good men, were vying for a front page byline. Editorship of a magazine was not as exciting, but Amy supplied an occasional adrenaline blast.

"Look, Amy, that's exactly why I chose you. Our readers aren't Ivory Tower types. They're simple people who expect us to "dumb it down" for them. Your lack of expertise in science would be an advantage. You'll ask the right questions in a halting sort of way, the scientists will take the cue, and they'll reply in simplified English. You're a good writer, and that's what counts. Also...," He broke a wry smile, "...It's an international story, so you'll get to travel. It'll be like a vacation on company time."

Amy pursed her lips in her standard, "I'm thinking," pose, then she relaxed her body, shrugged her shoulders, voiced a quick, "OK," and headed back to her desk.

"My trouble," she thought, "is that I'm married to my job." A divorce, however, was definitely out of the question. She was paid well for what she did, and job prospects in her field, or allied fields, were not that good. She became betrothed to journalism at a young age, as editor of the high school newspaper, and it was journalism in college. All in all, it was much better than most other majors.

Many girls from her era went to college, not to enter a profession, but to find a husband. Some of these girls gravitated to the easier majors, such as art, but others aspired to nursing. There, as a side benefit, they might later meet and marry a doctor. Amy started to think that this particular assignment would allow her to mix with some successful academic doctor types. Some of these must still be single, and maybe something will happen. Youth was still on her side. If not, she still had her first marriage to journalism to fall back on.

Amy had prior experience in science reporting in a series of articles she had written about biotechnology and genetic engineering. In that case, the emphasis of her articles was on the ethics of the research. There was a lot of research being done in biotechnology and genetic engineering, both by academics and companies looking to convert a few scientists' salaries into huge profits. Since the area was an ethics minefield, it wasn't that hard to write some lengthy and interesting articles on the topic.

One thing Amy learned from writing those articles is how valuable journalists are in educating the public. She was an educated woman, but she was surprised to learn about the things that were happening in biotechnology. Not only was there danger, but there was also greed. In the past, only criminals dabbled in both danger and greed, but now it was corporations.

Another thing she had learned was that scientists talk way too much, spouting what's indistinguishable from nonsense to the layman. You can't be a good listener. You need to interrupt often, and you need to lay sufficient track to get the story to station. Aside from the definite advantage that she had being a woman visiting a male-dominated environment, these scientists really wanted the public to know about the interesting things that they had discovered. A small problem was that interesting to them did not always translate into something being interesting in general.

Her bioethics articles were significant for another reason. She got thrown out of a pharmaceutical laboratory by asking the hard questions, like why they were selling their life-saving drug at $10,000 for a month's dose. She was young, then, and full of muckraking ambition. Now, it was more like a job, so she enjoyed the occasional perquisites of international travel.

As for now, her travel would be limited to the Internet, where she could research her topic. She knew as much about the message as any educated member of the public, but she needed much more. She needed to establish an accurate timeline of events, identify the principal characters; and, most importantly, link-up with the press agents at their various institutions.

The press agents, as usual, would be extremely cooperative, since their institutions value mentions in the press as a means to get more funding, more students willing to pay high tuitions, and so forth. Amy was fairly certain that the press agent's salary increases were based on the number and quality of press mentions, just as her own tenure was based on the number of letters and email messages about her articles, whether praising or damning, that were received by the magazine.

First, a quick trip to the bathroom, and then the coffee room. She intended to do her research in marathon mode, and those were her necessary prerequisites for two hours at her computer. That, and punching the code to send all her telephone calls to voice mail. Fortunately, she had a real office at the magazine, not one of those God-awful cubicles. She wouldn't be distracted by anyone else's ringing telephone or idle corridor gossip. Although she enjoyed watching the occasional sporting event on television, she saw no merit in Monday morning quarterbacking at the water cooler.

In reading the first background pieces, Amy realized how small a group these radio astronomers were. Just as the population of a species is dictated by the available forage, the limited funding for this field had limited the number of participants. A far smaller number of these had been searching for things like the SETI message, and many of them were just doing that part time.

Some dot-com millionaires had the vision to pump a little money into SETI research when the governments had seemingly abandoned it. Amy found it interesting that the public at large was actually footing the bill for such research through a small portion of the profits paid out from the purchase of their computers, cellphones and Internet. She was happy that the technology contingent of the one-percenters was doing more than just buying yachts and building huge mansions.

Before the discovery of the alien message, SETI was like a religion. Just like a religion, there were "Believers." There were also the "Non-Believers," who berated the Believers and thought that any dollar spent on SETI was wasted money. The Believers, however, were still fervent in their cause. They survived the funding pogroms, and continued on their crusade. It's not surprising that such a dedicated band of people would have succeeded in their cause. According to their calculations, it was just a matter of time and effort.

Aside from being celibate, the prevailing saint in all this, Blaise Tukey, didn't seem to fit the part. You would expect some flamboyant character, like a Richard Feynman, dropping sound bites at every step. Instead, you got a shy and older academic who seemed to resist the world's attempt at elevating him above his peers. He often credited his colleagues in the signal's discovery, and talked more about them and the signal than himself.

This led Amy to the topic of the first article in her series of SETI signal articles. She would write about how the science, itself, always takes center stage in the practice of science. What Nature says is always more important than what a particular scientist says. Nature's words are so important that you need to hear them first hand, and not just believe another scientist's data without verification. Research results need to be replicated. That was more easily done with the SETI signal, since you just needed to point your existing radio telescope in the right direction and listen.

The senders of the signal must be scientists, themselves, so they might share the attributes of their Earthling counterparts. Amy imagined the coffee room conversation.

"Did you hear that *Old Brown Tentacles* has been sending messages into space? Not only that, but he's using radio! How retro!"

In her reading, she discovered some reasons why an advanced civilization might send signals into space. There might be civilizations free from want in which all labor is done by robots. The space aliens, as a way to assert relevance or relieve themselves from boredom, would do interesting projects like sending signals into space as a hobby. Their budgets might be potentially unlimited, so a powerful space transmitter could be built on the whim of a an individual using a robot workforce.

As the initial dose of coffee started to diminish in effectiveness, Amy started to daydream about what she would do with a robot workforce and unlimited resources. A step up to a nice house from her present apartment would be nice. After that, it would be hard to decide. She wondered whether any of her robots would be anatomically-correct. In that case, she might not feel the need to leave her house at all, at least for a few decades.

That led to one obvious reason why the alien message was important. It meant that there was at least one civilization that survived long enough to do such things. At the least, it gave hope that humans, too, might reach that point. There might be a technological tipping point, somewhere in the future, past which the extended existence of humankind was assured.

Amy was reminded of the utopian and dystopian novels that she was forced to read as a high school student. Modern life seemed to have developed as a mixture of these themes. The government had become more totalitarian, enabled by surveillance technologies, while the lives of their citizens had become incredibly enriched through the instant communication capabilities of cellphones and the Internet.

So, there she was, sitting in a comfy chair, "working." Centuries earlier, sitting in a comfy chair would not have been recognized as work.

18 RECRUITING ROOS

Roos was startled from his nap by the ringing telephone. He didn't like telephones, but he admitted their necessity. Solitude was one thing, but isolation was a different story. He needed some lifeline to the outside world, he still had family, and he was in the egg business. Very few people knew his number, but even then he still screened all calls through his caller ID box. His answering machine received those he chose not to deal with immediately. He was likely the only person on earth with no e-mail address, but he did have an Internet connection. Since his eyes were too unfocused to read the name on the caller ID box, he decided to answer.

"Roos here."

There was a lot of background noise, but the voice was clear.

"Martin, this is Joe DiGiuseppe, how have you been?"

Joe DiGiuseppe! Roos hadn't seen Joe in years, at least as long ago as his move to the chicken farm.

"Don't get me wrong, Joe, I'm happy you called, and all that... but, how did you get my number?"

"Martin, you forget that I work for the NSA. We've got *everybody's* number!"

Joe had been a classmate of Roos in graduate school, and they had completed their Ph.D.s at about the same time. Joe's research interest was algebraic number theory, quite a bit removed from Roos' differential geometry, but they had been good friends all through school, swapping teaching assignments when the other was in crunch mode. The National Security Agency employs about half the mathematicians in the United States. Joe and Martin went their separate ways in the 50/50 split.

"What have you been up to, Joe?" I don't read the journals anymore, but I suppose whatever you're doing isn't published."

"I'm just an administrator. Math is a young man's game." He regretted saying that. He didn't know Roos' present mental state, and he definitely didn't want to offend him. "But my boys at the Agency keep me busy; and, more importantly, they keep my mind sharp."

"Joe, you were always a player in the big leagues. I was sorry when you sold out to the NSA. So, why the call?"

"Well, here's the cloak and dagger part... I can't tell you over the phone. If you allow me a visit, we can talk at your place. I'm about ten minutes away."

Roos surmised that Joe was calling from a cellphone in transit to his place. No directions required for the NSA.

"Sure. I'll put on some coffee."

"Great. We thought we would land in the field about 100 yards east of the silo. Would that disturb your chickens too much?"

Roos had to think for a second. Silo? Must have gotten that from satellite surveillance photos, maybe in real time. Land?

"You're in a helicopter?"

"Martin, it's the only way to fly!"

"The field's OK."

"See you soon."

Roos was standing alongside the silo to greet DiGiuseppe when he arrived. He was wary of the helicopter blades, so he stayed in place. The noise of the chopper was disturbing to a man accustomed to the quiet of his rural life. Joe opened the door, shouted a few instructions to the pilot, jumped the few feet to the ground, and walked towards Roos. The pilot throttled down the engine, and remained in his seat. This was going to be a one-on-one conversation. There were the usual pleasantries as they walked to the house - nice place you have here, I just heard that so-and-so remarried, etc. - but Joe was all business as they grabbed their coffee cups and sat at the kitchen table.

"Martin, you've likely guessed that this is not a social visit. I'll get right to the point. It's about the LGM."

"LGM? Oh, you mean the SETI signal."

"Yeah. Once the media got hold of the Little Green Men catch phrase, there was no stopping it. It had some history behind it, and thus a certain credibility, so the phrase just took off. They write LGM now for short, but it could just as well have been BEM, for Bug Eyed Monsters."

"But what do the Little Green Men have to do with me?"

"Martin, a lot has happened in the last few weeks. It would have been easier if the government had discovered the signal, not a private foundation, then we could have kept the discovery a secret until we understood it. Many years ago the government decided that SETI was not a useful way to spend its money. That little bit of money we saved at that time pales in comparison with the money we've spent in the last few weeks trying to bring the government back into the game. No one knows at this point whether we have a message, or just a signal beacon, but if there is a message, national security requires that we have a copy. It would have been nice to have the only copy, but that's not possible at this point."

"Joe, there's a big distinction between a signal and a message. Almost anyone can record the signal, but decoding a message is another story. You could be first to decode, and that might be an advantage."

"We thought about that, but the stakes are too high. We have no measure as to whether we have an advantage here over other governments. Something like knowledge of the Chinese language, or at least thinking like a Chinese speaker, might offer better clues to decoding. Also, there's this whole thing about the SETI protocols. The protocols aren't international law, but the astronomers consider them to be sacrosanct, and we would take a lot of flack if we decided to do things on our own."

A quick slurp of coffee, and then he continued.

"There are also the fringe groups, religious zealots, and anyone with a distrust of government. That's why the President will give a speech this evening establishing an International LGM Project and inviting all nations to participate. We're able to steal the initiative here, since the discovery was made in our country, and our NASA tracking stations have been getting continuous recordings of the signal since shortly after the first press conference. That's something most other governments can't do. This is where you come in. We are proposing you as the Director of the LGM project"

Roos dripped a bit of coffee onto his lap. Fortunately, he liked his coffee cold, with a lot of milk.

"Martin, you're the perfect choice. Your technical credentials in cryptanalysis are superb; but, more importantly..." As if for emphasis, DiGiuseppe surveyed the chicken coops through the picture window. "You're your own man. No one will claim that we've got you under our thumb. You're the perfect choice for this venture. I trust you, and I've convinced those in charge that you can be trusted."

It was hard for Roos not to be a little excited. At the same time, his mind dredged up all the memories and reasons why he retired to his chicken farm in the first place. But now, after all those years, he had to admit that he was getting a little bored. This was a chance of a lifetime! He could try it for a while, just to ease the boredom, then come back here when the responsibility outweighed the novelty.

"I must say that I'm interested. But who will handle the farm?"

DiGiuseppe laughed, "We'll have the Secretary of Agriculture tend your chickens! With what's at stake, that's the least of our worries! So, what do you say?"

Roos took another sip of coffee, decided what the parameters of a logical decision should be, and he began to calculate. Others might have wondered about the trance-like state that Roos had entered, but DiGiuseppe was familiar with these. They rarely lasted more than thirty seconds, although he was inclined to allow Roos a full minute because it was a difficult question.

Roos' mind must still be sharp. His response came in less than half a minute.

"Yes, I'll do it. But, Joe, you need to promise one thing."

"What's that?"

"You need to make it easy for me. I'm still a recluse at heart."

Before Joe departed, he outlined how Roos' staff would be organized; and, with the many available staff, how the routine matters would be handled without any effort on his part. Essentially, he would be paid to sit and think. It was just a thumbnail sketch, since the helicopter was waiting, but Roos was satisfied.

Roos was outlining in his head his own vision of his new career, which was essentially just an updated version of his old career. It was as if the aging professor was finally awarded a chair, so he could sit back, think, and do those wonderful things that made his other colleagues famous. Judging from what DiGiuseppe said, he would emerge as the most famous of them all.

Did he really desire fame? He had retired to the chicken farm to get away from it all, and now he was diving back into the hectic world he had left. Perhaps he had subconsciously realized that there was never any way to escape civilization. As a chicken rancher, Roos still needed to mind the books, maintain connections with his upstream processors, his hired help, and so on. His new position was starting to sound more like a vacation than work.

His sleep came slowly, since his active mind was organizing its thoughts around his new role in life. One small problem was that he didn't have a car. He always hated driving, and his hired help handled all the transportation needs of running a farm. He would have let his driver's license lapse, were it not for the fact that those had become *de facto* identification cards. DiGiuseppe would need to supply him access to a limo and driver.

After lying in bed without sleeping for nearly an hour, Roos arose and walked out to his front porch. The night air was cool, but not uncomfortable, and the sky was clear. He looked up, wondering which of those many stars the LGM called home, and then realizing that their star was likely too distant to be visible to his unaided eye.

The star was possibly invisible to all but the most powerful telescopes, and it might even be on the other side of the Earth at that time. Wherever these LGM were, their message was already affecting the lives of people on Earth, although the content of the message was still unknown.

19 PRIME LUNATIC

Jake was the Prime Lunatic. At least, that was his appellation in the media. Because of its negative connotation, the name hurt him deeply, but he would never admit it to his compatriots. Since he shunned the mainstream media web sites and television news programs, he rarely heard the term. His brethren knew what he was all about, and that's all that mattered.

The whole lunatic thing could have been avoided by a simple name change at the start. Since most of the members were technologists and engineers who wished to exchange ideas, it was suggested that they name themselves after the famous Lunar Society of the Industrial Revolution. Now, there was a bunch of enlightened individuals!

The Lunar Society, first called the Lunar Circle and renamed after its first decade, was a group of intellectuals who met once a month at a member's house for dinner and conversation. The lunar part of the name came from the tradition of having the meeting at the full moon, to make for easier travel in those dark days before street lighting. The members referred to themselves as *Lunaticks*, but never as *Lunatics*.

The Lunar Society was an assemblage of the most notable men of the time. It included Erasmus Darwin, the grandfather of Charles Darwin, and Josiah Wedgwood. Benjamin Franklin occasionally visited. Wedgwood was Jake's favorite of the crowd, since Wedgwood was as close as you could get to being a technologist in those days. Jake's formal occupation, what he referred to as his "day job," was as an engineer.

Wedgwood appealed to Jake at another level. Wedgwood was an early abolitionist who proudly wore a medallion of the Society for the Abolition of the Slave Trade. The medallion had an image of a chained slave, and it was emblazoned with the motto, "Am I Not a Man And a Brother?" Wedgwood donated funds to have many of these medallions made.

Unlike the *real lunatics*, as Jake called them, his group of Lunatics was not anti-science. They just wanted science to be more directed towards human ends. Along with that creed came their anti-military sentiment, since money could be better spent on people than weapons of war.

As soon as The Message was detected, the Lunatics wondered what it meant for Mankind. Since they were knowledgeable people, they didn't subscribe to any of the Evil Alien hypotheses. As the online conversations blossomed, one opinion stood out. We were humans, dammit, and we didn't need any alien influence in how we conduct our business!

Sure, there might be some wonderful things in the message, like a cure for cancer, but would any of that be worth giving up our essential humanity, a work in progress that's been ongoing on this planet for millions of years? The message wasn't evil, *per se,* but it would be an unfortunate contamination of our essential human character. The Lunatics were against the message.

There seemed to be an arms race developing around decoding the message, perhaps for some advantage of one country over another. It was the typical human reaction, and one that argued against even the reception of the message, let alone its decoding. Fat chance they could stop it through any powers of persuasion, since The Message had become an industry unto itself with thousands of stakeholders.

That's why several members of the group had started brainstorming more effective means of stopping The Message. Sit-ins and pickets were surely not that effective. They would need to think big. Jake did not like violence. If they resorted to violence, they would have sunk to the level of the military that they detested. Jake had made it clear that whatever they did to stop The Message should be non-violent, but effective.

What that might be was not clear at this point, but his people were a tech-savvy bunch. Since it was apparent that computers would be involved in the decoding process, some members were working on virus code. This seemed like a really good idea, so Jake approved the virus work. Sure, it was illegal, but at least it was non-violent. And, if they did it right, it could never be traced back to his organization - If you could call such a loosely knit group of people with a common interest an organization.

Unlike the professionals that they were pitted against, this was just part-time work for his Lunatics, who had day jobs. Jake had once heard the slogan, "Work smarter, not harder," and he tried to keep his Lunatics on this track. They joked that it wouldn't be all that hard, since government bureaucrats were never known to work harder, or smarter.

Jake always worked towards a consensus for action. When you have a volunteer organization, it's not good to antagonize the troops. They would just defect, perhaps starting a competing cell. Jake saw similar things in the development of open source software, where groups sometimes "fork" the software development in different directions. Sometimes the forked distributions became more popular than the original.

Since the Lunatics seemed to be more organized than the keepers of The Message, there wasn't any urgency for action. Jake and his Lunatic associates were in a holding pattern until there was an actual target for attack.

Jake's life story had prepared him for the Prime Lunatic position. He was a frail child, likely a result of his mother's lack of proper nutrition and medical care when she was pregnant with him. She also smoked. As a consequence, he wasn't very athletic, so he didn't mingle much with the other boys. His was born into a lower middle class family, and their lack of money isolated him even more.

Since there were few distractions in his life, Jake did modestly well in school. He wasn't at the top tier of his class, but he was close. He suspected that a little more money for books and other study aids might have given him access to the very top. For that reason, he had developed a resentment for the students at the top. They got there through an unfair advantage.

While in high school, Jake discovered the late night AM radio talk shows. He had always known that there were some things not quite right in the world, and he was learning about the conspiracies perpetrated by the government and wealthy people. He never called into those shows, but many who had seemed to be talking about the same injustices that he, himself, was experiencing. His sadness about his life had turned into a simmering anger.

After high school graduation, when many of his classmates were headed to prominent four year colleges, Jake was at the community college. He needed a part time job to pay the tuition, and this cut into his study time. Fortunately, his fellow students were not much competition, and his state had a plan that guaranteed admission to a state college upon successful completion of a community college degree. His stellar performance at the community college also got him a much needed scholarship.

Even with the scholarship, Jake needed a part-time job, so he didn't have any time for socialization after classes. In class, the girls chose to sit near the athletic boys. He had considered inviting one chatty girl to a movie on one of his rare nights off, but he had waited too long. She had taken to walking arm-in-arm with another boy into class.

Jake received an electrical engineering degree with a high grade point average, only to find that engineering, in general, wasn't a respected profession; and, that an engineering degree from a state school was highly discounted. He ended up in a Dilbertesque cubicle within a sea of other engineers, doing fifty hours of work for forty hours of pay.

Jake spent nearly all his non-working hours at his computer. The Internet was a world that was far friendlier to him than the one outside his apartment. After he had started posting his opinions on various message boards, he put his computer skills to good use by writing a program that scanned his favorite message boards for key words. In that way, Jake was instantly signaled about new posts in the topic areas that concerned him the most. By keeping on top of these topics and responding promptly, Jake had become a celebrity in a small circle of like-minded people. He had a following.

While work was the defining principle for many people, Jake's work became just something that he needed to do in order to pay the bills. Although he completed his work assignments on time, he was mentally detached from work. There was no extra effort for an employer who didn't reward its workers sufficiently. The Internet became his life. Work was just a nuisance.

Often, rushing to his computer after arriving home from work, he would get so involved that it was nearly midnight before he realized that he hadn't eaten. He had reduced his sleep time to a mere six hours each night, telling himself that he would get more rest on the weekends, but that never happened. If it weren't for some very strong coffee each morning, he would have found himself in a ditch at the side of the road after falling asleep on his morning commute.

Many of his message board correspondents suggested a simple solution to his sleep problem – Various stimulant drugs, all easily obtained through back-channel sources on the Internet. Although not strictly a puritan, Jake was definitely anti-drug. When he was awake, he wanted to be fully sensible. And so, there he was, night after night, at his keyboard.

20 SIGNAL ANALYSIS

Should credible evidence of extraterrestrial intelligence be discovered, an international committee of scientists and other experts should be established to serve as a focal point for continuing analysis of all observational evidence collected in the aftermath of the discovery, and also to provide advice on the release of information to the public.

– The SETI Protocols

Things moved quickly after the President's speech, and it seemed that money was no object. Joe DiGiuseppe was acting as the government liaison. With purchasing card in hand, he was handling most of the boilerplate. It was decided that there would be several meetings of the US principals over a span of two weeks, before the first meeting of the international LGM committee. An invitation list had been easy to compile, owing to the close-knit nature of the SETI community.

Roos' mind was churning in the background while all this was happening. Roos did not need a meeting to do serious thinking; in fact, he worked best in solitude. If he were to beam a message to the stars, one he wanted to be understood by any species intelligent enough to intercept it, how would he do it?

Scientists had been playing with this idea for more than fifty years. Some thought that a picture would speak a thousand words and proposed sending simple images of things like the periodic table to establish a frame of reference. Then they would send images of our solar system, and, eventually, images of ourselves. The Pioneer 10 spacecraft, the first earth object to leave the confines of our solar system, contained a plaque with such images, including a naked man and woman. Roos laughed when he remembered that the biggest controversy about the naked couple was not their nakedness, but whether or not they should hold hands. There was fear that an alien viewer might interpret a couple holding hands as a single organism! Images were all a waste of time. Roos knew that mathematics is the universal language.

The most universal feature of mathematics is counting, and the idea that two is more than one. The prime numbers would be key to any message. In fact, the early images intended for aliens were dots arrayed in prime number columns and rows. When given a certain number of dots, say 2675411, any alien mathematician would know that this was the product of 1433 and 1867. He would then arrange the dots in 1433 rows of 1867 columns and see nothing interesting. Then he would laugh (in his alien sort of way) and try 1867 rows of 1433 columns, which would give him the encoded image. A nice idea, but a very anthropomorphic idea that assumes that aliens have eyes. Perhaps they sniff their way through life. Of course, they must have some sort of vision, even if it is based on scent, so the image might be understandable.

Once numbers are established, it's easy to move on to geometry via the 3-4-5 Pythagorean triangle and the value of pi. From that point you can map your solar system using spheres of fixed dimensions at various distances from each other. Now here's the tricky part. What do you use for a length scale? The obvious choice is the wavelength of your radio signal, but there are others. For very small distances, the Plank length would be ideal. For intermediate distances, the distance between two hydrogen atoms in hydrogen gas, the major component of the universe, could be used.

Once the spatial extent of objects and their placement is established, you can map molecules, finally arriving at DNA. If DNA is common to life in the universe, then we could manufacture our own alien to see what they look like. Here's where the ethical questions enter, so it's no wonder that a goodly number of philosophers and theologians were invited to the meetings. Of course, Roos was not so naive to think that governments wouldn't be doing such things in clandestine laboratories no matter what the theologians decided. He sensed the plot of a good science fiction movie coming on, more fodder for the Lunatic Fringe.

All Roos' thinking was based on the idea that the message was meant to be intercepted and understood. There was the chance that we were just eavesdropping on some interstellar chatter, maybe even an extended grocery list. The signal may even be an accident or some sort of radar beam, something not intended to communicate at all.

The first meeting was to be held in a hotel in Zurich, a reasonably neutral locale with all the amenities of the civilized world. The press applauded this decision, since the logistics of telecommunication couldn't have been easier. Joe DiGiuseppe and the US State Department had tried to limit the meeting to two hundred participants, but every nation in the world wanted representation, and the attendance list blossomed. Roos was afraid that the technical discussions would be buried under pretense and diplomacy, and he proposed a plan that was eventually accepted.

There would be two parallel sessions, one for the true technical issues and limited to about two hundred scientists, and the other would be an issues session at which all the politicos could posture, blow hot air, argue about finances, etc. Roos, for one, was happy to quarantine the bad seed, but there was a need for a plenary session, an opening ceremony to introduce the imputed leadership and keep the press happy.

The opening ceremony was structured in a way to establish the US lead in the venture in such a logical progression that there could be no argument. The President of the European Council would acknowledge the presence of various dignitaries, and he would then introduce Blaise Tukey to thunderous applause. Tukey would acknowledge the assistance of NASA and other US government agencies, give a short speech, wait through more thunderous applause, and then introduce Roos. Roos would outline the agenda of the meeting and invite everyone to lunch. At least, that was the plan.

⊢⊢⊢⊣⊣⊣

Amy enjoyed being anywhere else than at her office, so she enjoyed her travel assignments. She also admitted freely that nearly every restaurant chef could cook better than she, so she didn't mind being away from home as well.

The problem with travel was getting there. Some light reading helped to alleviate the boredom of flight, but she hated the airport security. This was a greater problem with international travel, since everyone was marked as a terrorist until proven innocent, and this was an affront to her sensibility. With her volatile personality, she had the additionally difficult task to suffer fools gladly.

Somehow, she arrived at the conference hotel unscathed. The press officers for The Message were extremely helpful and attentive to detail. There was a van to shuttle arriving journalists from the airport to the conference hotel. There was even a hotel representative in the press room for check-in. She was given her press credentials, a thick information folder, and an invitation to a cocktail hour with some of the minor players in the message organization.

Her only exposure to the principal players in the message organization would be at a press conference. This was the most efficient way to do this. A one-on-one for each of them with every journalist would have taken a week, and the journalists would have all asked the same questions, anyways.

Amy didn't drink, since there were dangers in being a drunk woman in a den of male journalists, but she did plan to attend the cocktail party. The hors d'oeuvres were always excellent, and while the others were queued at the bar for drinks, she could corner some of the newsworthy people while they were still fresh and talkative. She set her cellphone alarm for an hour before cocktails and lay down for a short nap.

⊢⊢⊢⊣⊣⊣

When Roos arrived at the hotel the day before the start of the meeting, things started to look a little more complicated than expected. First, the Lunatic Fringe had encamped itself around the hotel. The swiss police had tried to establish a Lunatic-free zone for several blocks around the hotel, but the crush of people was so strong that the perimeter was eventually reduced to a block. This was an adequate compromise for the demonstrators, since they were keen on being able to see the hotel, but not much more. Some carried signs and hoped for contact with the press, but they seemed peaceful enough. It was more like a party atmosphere, an LGM Woodstock, and their only impact on the meeting was to impede traffic.

There was a message from Joe Digiuseppe waiting for him at the check-in desk. Roos was invited to dinner with Joe and Blaise. They would meet in the lobby at 6:00 PM, which gave Roos about two hours to unpack, shower, and check the television headlines. He was distracted for a while by the incongruity of an I Love Lucy episode dubbed in German, and before he knew it, it was dinner time. He reached the lobby first, cozied himself into an overstuffed chair, and stared out the window. The police were stationed about fifteen feet apart around the perimeter of the hotel. Behind each policeman was a large duffel that likely contained riot gear and an assortment of non-lethal weapons. The protesters were nowhere to be seen, since they were nicely blocked by landscaping and other buildings.

As Roos had expected, Tukey arrived before DiGiuseppe. Technical people are traditionally punctual, whereas the management types are notoriously laid back. Joe should have scheduled their meeting as "sixish," instead of 6:00 PM. The two were a few minutes into casual chit-chat when Joe Digiuseppe arrived, accompanied by a beautiful and well dressed woman introduced as Vanessa. Vanessa, as Roos had heard, was Joe's Significant Other. They weren't married, but they had lived together for several years, a period longer than many marriages.

Vanessa was an impressive sight, and Roos could see that Tukey was genuinely interested, quite a bit more interested than Roos was himself. It wasn't that Roos was disinterested in women - he was a confirmed heterosexual - he just didn't have time for *The Chase*. Courtship takes too much time, and women demand too much attention. Also, Roos' typical pose of staring blankly into space while contemplating some math problem was not an endearing trait to most women.

There was another, darker reason for Roos' aversion to women. Roos didn't like hair. Hair on a woman's head was fine. It was the crowning glory of her beauty, but hair anywhere else seemed to be unfeminine. Perhaps it was the Grecco-Roman statuary he viewed when he was young, but he never expected women to have hair, there. It was quite a shock when he found out in his college days, and the shock lived inside him all these years. Maybe it was a quirk of his mathematical mind - the whole thing just didn't add up - but it was as real for him as any proven theorem. Since it was not polite to inquire about a woman's bikini waxing practices on a first date, Roos was essentially excluded from this aspect of an adult social life.

Joe had a rented car, and from the conversation during the short trip to the restaurant, Roos surmised that Vanessa was a non-technical type. He was pleased when his hypothesis was confirmed by her revelation that she was a lawyer employed by the NSA. That was obviously where she and Joe had met, and Roos could see how Joe had been attracted to her. She was non-technical, but she had the sparkle and demeanor you can expect to find only in the higher circles of the intelligentsia; and she could hold her own in any conversation.

When they arrived at the restaurant, parking turned out to be a problem. They had to park about a block and a half from the restaurant and walk, prompting Blaise to quip, "If we were in a movie or a television show, there would be a parking space for us right at the front door of the restaurant, so I guess this is real life." However, there was no wait at the restaurant, either because of a reservation, or some under-the-table maneuvering by Joe.

The meal was a perfect mix of international cuisine and light conversation. Martin and Joe were able to catch up on their respective histories since graduate school, while Blaise and Vanessa listened with either rapt attention, or nicely veiled boredom. Martin caught Blaise several times looking longingly at Vanessa, much like Eve would admire the forbidden fruit. While they were waiting for dessert, Joe spied some people entering the restaurant, and excused himself.

"He'll be gone for some time, said Vanessa. "I recognize those people. They demand a lot of attention. That gives me a chance to ask for a layman's version of The Message and what we hope to accomplish at this meeting, and later."

Blaise began.

"Well, from a scientific standpoint, it's certain that The Message is genuine; namely, that we've actually discovered the Little Green Men. Our analysis of the LGM signal and the signal from the associated pulsar show that the LGM are very far away. They're so far away that any physical visitation would be impossible, either their visiting us, or vice versa. That's assuming that the speed limit on travel, the speed of light, is actually the case. I'm too much of a scientist not to believe that some parts of Relativity might someday be shot full holes, so this may not be the case. It may happen that Relativity is just a very good approximation, but not the entire truth, just like Einstein overthrew Newton in some sense when he discovered Relativity. So it may be that the speed of light is not a universal speed limit for space ships, but as far as we know, it is."

Blaise sipped his water, and then continued his lecture.

"The signal may even be a guidance beacon for some sort of space ship on a course to Earth. That's what some of the Fringe believes. As for the Message from God Fringe, God would have done things much more easily. Other fringe elements think that The Message contains answers to all the world's problems, and it will lead to immortality through medical science. That would be nice, and I welcome that as much as anyone. But if the content of The Message is nothing more than what we already know, The Message is still important, since it tells us that someone else is out there."

Vanessa joined the conversation.

"I remember when the government stopped funding SETI, and the private foundations had to take over. The politicians argued that you guys had already looked for decades and didn't find anything. The SETI response was that if we looked for a very long time and didn't find anything, that would be just as important a discovery – that no one is out there, that we're alone in the universe. Martin, what are your thoughts about The Message and how to decipher it?"

Martin had been as attentive to the conversation as he could be, which in his case was usually one out of two ears. He always had some background process running in his mind that demanded at least 50% brain time. This time he had heard and understood the question the first time, probably because of the sweet voice behind it.

"It's interesting you should use the word decipher, rather than translate, as the media do. That's the way I'm seeing it, as a code to be broken. Cryptography, or the creation of ciphers, is relatively easy. Cryptanalysis - the breaking of ciphers – is what's hard. One man, Arthur Scherbius, invented the Enigma cipher machine used by the Germans in World War II. But it took hundreds of people to crack the messages using this cipher. I think we're aided by the idea that the authors of the LGM message, the LGM themselves, want to have The Message understood. I also believe that they're using a universal key, mathematics, to aid our analysis. At the same time, we need to remember that they're not human - they probably don't think like we do, don't have the same senses, the same desires or motivations. That's what will make this hard, that there are few common reference points."

Tukey interjected, "I must disagree. The energy levels of electrons in atoms are the same everywhere in the Universe. The Periodic Table of the elements is the same. Our views of the stars are the same, but from a different perspective. K-meson decay even gives us a way to tell left from right. We have many reference points in areas that really matter."

"Spoken like a true left-brainer," said Vanessa. Everyone laughed, including Tukey, and Vanessa continued.

"There's another interesting aspect of the LGM which has excited more than one faction of the Fringe. That's the idea that we are the children of the LGM - that they purposely seeded life here, or they genetically engineered the apes into intelligent humans."

"Part of that is the panspermia hypothesis," replied Tukey, "That life originated in only one place at one time in the distant past, and it somehow propagated throughout the universe. There are two aspects of panspermia. There's passive panspermia where microbes are knocked off planets by meteor impacts, and they somehow survive a trip through space to land on another planet. That idea is not that far fetched, since spores have been known to survive extreme conditions. Then there's the other aspect, active panspermia, when the spacemen seed life on other planets. It might be as simple as leaving behind some garbage, or more purposeful like genetically engineering organisms to thrive on other planets. They may think it's something like God's Will that they undertake the task of populating the universe with life. Something like this would make the origin of life on earth easier to explain."

Roos, who had been uncharacteristically attentive, chimed in, "That may be true, but that begs the question of where the original civilization come from. Mathematicians like to tell this little anecdote about Bertram Russell. Russell had just given a lecture about the solar system, and a member of the audience admonished him that he had gotten something wrong. The sun was carried on the back of a turtle. Russell asked what carries the turtle. It's simple, it's turtles all the way down."

Once again, there was laughter all around. At that point, Joe DiGiuseppe appeared.

"So, what's the joke?"

"I was telling the gentlemen about the time you proposed to me," said Vanessa. "I thought you were so sweet, but they burst into laughter."

"You told them what!"

"Don't worry Joe, said Martin, "Vanessa didn't tell us anything of the sort, but now that she's mentioned it, my interest is piqued. After all, you aren't married, so she must have refused."

Vanessa laughed. "I have my own reasons, but the story's sweet just the same. It's just one story you'll probably never hear."

Joe seemed relieved.

After a good dinner, it was agreed that the hour was late, and everyone needed rest before the morning. Martin and Blaise made plans to meet in the lobby the next morning for breakfast. Back in his room, Tukey had a hard time sleeping. He wasn't thinking about his address in the morning. He was thinking about Vanessa. Perhaps a geek's life was not as rewarding as he had once thought.

21 INTERNATIONAL COOPERATION

Should credible evidence of extraterrestrial intelligence be discovered, an international committee of scientists and other experts should be established to serve as a focal point for continuing analysis of all observational evidence collected in the aftermath of the discovery, and also to provide advice on the release of information to the public.

– The SETI Protocols

Blaise awoke the next morning, not to the sound of the alarm clock, but to the chants of the demonstrators. It wasn't just the birds who arose with the sun. Most of the demonstrators had camped around the hotel in sleeping bags, and with the rising sun they started to wander the streets looking for the porta-potties the city had so presciently provided, and food. Blaise flipped on the television and watched the screen reflected in the bathroom mirror as he shaved. The report was upbeat. No violence and no threat of violence. Blaise showered and took the elevator to the lobby for breakfast. Martin had phoned to say he wouldn't be available for breakfast, so Blaise just grabbed a quick cup of the lobby coffee to drink as he scanned the newspapers. He was somewhat pleased to find his photograph on one of the inner pages of USA Today, although it was the same dated photo his press office had been distributing for weeks. He was happy that he had uncharacteristically had his hair styled for the occasion.

The gradual onslaught of delegates through the lobby door was his cue to head backstage. If any stopped to chat, he would be obligated to enter a conversation, and that small conversational group would become larger by the minute, and he would be trapped. He needed some quiet time to compose himself before his speech.

Tukey had given many presentations to colleagues at scientific meetings, but in those cases there were just a few tens of people in the room, a hundred at most at some international meetings. Those presentations were unscripted, and his slides were his memory aid as to what words went where.

Here, he would be talking to a huge audience in a venue quite different than the hotel meeting rooms filled with folding chairs and reeking with the smell of too many cups of inferior hotel coffee.

Also, his speech was scripted. Not only that, but it was edited to the point at which he hardly recognized his own words. A team of people had considered the consequences of every word and its ability to offend even the farthest fringe group.

One point that had been pounded into Tukey's head was that he was not to say anything about the senders of The

Message. The word, *people*, to describe the senders was expressly prohibited.

After a time unnecessarily reviewing his notes, Tukey was escorted to the podium ante area to watch the lead-in to his speech. Speakers, apparently, were not expected to sit while they were waiting, although Tukey yearned for a chair. Instead, he leaned against a wall and relaxed as best he could.

The President of the European Council was the first to speak. He acknowledged the presence of various dignitaries, all to thunderous applause. Tukey wondered whether carpal tunnel syndrome was a common complaint among older diplomats after years of such meetings.

Blaise Tukey was introduced, and he went through his script, all nicely displayed on a Teleprompter so he wouldn't mistakenly use the wrong word. He acknowledged the assistance of NASA and other US government agencies, gave the barest history of his discovery, and wrapped up in five minutes. Brian's name was not mentioned, although Tukey had insisted that a footnote reference to their joint paper announcing the discovery be included in the conference proceedings.

Just as Tukey was wrapping up, he noticed a lot of activity at the back of the auditorium. The doors, which had been closed at the start of the ceremony, were now somewhat open, and they were swinging in and out, as if someone was not quite certain whether to enter.

The doors suddenly burst open, and Tukey could see that the swinging doors had been backed by a line of police attempting to withhold a crowd of protesters. As he had learned later, some protesters had entered the hotel though a side entrance propped open by a hotel employee of kindred spirit.

The protesters, chanting slogans, could not be contained by the police, and they started moving down the central aisle. They didn't reach Tukey and the podium, since another squadron of guards emerged from the back of the stage, ran down the main aisle, and stopped them in their tracks.

It was not a violent protest, but it did make for good press coverage, so it was a success in the eyes of the protesters. It wasn't clear to Tukey what was being protested – Perhaps it wasn't related to The Message at all.

The protest was wrapped up in a little more than a minute, but it took much longer for the delegates to settle down. All the while, Tukey stood at the podium waiting for an appropriate point to introduce Roos.

Roos, whose professional mode of operation was to sit in solitude for hours at a time, pencil in hand, alone with his thoughts, was pacing backstage apprehensively. Roos, who was uncomfortable being in a group of more than a few people, was now faced with the task of addressing a room filled with a thousand. He convinced himself that he just needed to mouth the words on the Teleprompter to get through.

After his introduction, Roos walked across stage to the podium. He was somewhat shocked by the loud and lengthy ovation he was given. He wasn't anyone *that* special, and he hadn't even started to prove himself in his new job. Usually, an ovation like that would bolster a person's confidence, but for Roos it had the opposite effect. They had a high expectation of his skill. What if he failed?

Fortunately, Roos' speech was more functional than inspirational. He outlined the agenda of the meeting, specifically pointing out the separate technical track that he and Tukey would co-chair, and he invited everyone to lunch.

Roos thought that meetings, even monumental ones of this sort, were always tedious, but he was happy that some important things got done. A principal milestone in his mind was the formation of the various working groups, both technical and administrative, by which the message organization would function after the meeting.

Roos' initial sketches of the most logical type of organization had been converted to organizational charts that were populated with names at the meeting. Since it was obvious that there might be more to the physical nature of The Message, as opposed to its supposed content, and a need to ensure the best methods of detection, Tukey became chairman of a Reception Research Group.

Roos, who was also the "chair of the chairs," took the added responsibility of being chairman of the Message Decoding Group. Other technical groups involved maintenance of the radio telescopes used for reception of The Message, the network infrastructure for dissemination of The Message worldwide, and in archiving The Message and the organization's activities for future generations. Roos paid no attention to the non-technical groups. In his mind, their only function was to supply money for technical purposes.

⊢⊢⊢⊣⊣⊣

Aside from a few insights at the closing press conference, Amy hadn't discovered anything she could write about that would differentiate her from the many other journalists at the conference. A lot of follow-up work would be required, perhaps first visiting one of the radio telescopes, then the location of Message Headquarters, wherever that might be sited. Since her assignment was near-term, she wasn't counting on being hosted at Message Headquarters, since that likely wouldn't be established for a year.

Now that the conference was over, she was in wind-down mode, which meant a relaxing meal before her next day's flight. She was looking for genuine food in a real restaurant, not the bland cuisine in the stilted restaurants around the hotel. She knew what she wanted, and those types of eateries were located in the university district, no matter what city she was in.

Roos was likewise winding-down, so he accepted an invitation from Tukey to join him and some colleagues for dinner. Roos had declined similar offers from some of the non-technical attendees, but he felt a stronger kinship to technical types, so he accepted Tukey's invitation. Since these were also academics, they selected a Middle-Eastern restaurant near the university.

Amy found the "Ali Baba" without too much trouble. As she sat at the bar, beer in hand, waiting for a table, she thought about the incongruity of it all; that she, a woman, would be drinking alcohol in a Middle-Eastern restaurant. Just then, a boisterous crowd of middle-aged men, easily marked as academics, entered the restaurant. Since this was a large group, they had been able to reserve a table, so they were seated immediately.

Amy suddenly realized that in their company were Blaise Tukey and Martin Roos! This was an opportunity too good to bypass!

She hadn't gotten ahead in journalism by being timid, so she approached the table where the men were just getting settled. There were no women in the group, which gave her an immediate advantage. Addressing no one in particular, while exposing her conference ID badge, she said,

"Gentlemen, may I join you?"

Tukey, who was the putative leader of the group, saw no reason to object, and he brought her a chair from another table. Roos was shielded my many bodies at the back of the table, but she was seated across from Tukey. She introduced herself to those nearby, sipped her beer, and studied the menu. Amy had learned that it wasn't good form to start questioning too early, so she participated in idle chit-chat until all orders were placed.

Since Tukey had invited her to their table, Amy directed her first question to him.

"Doctor Tukey..."

"Just call me Blaise."

"OK. Blaise, as we've heard at the conference, it's far easier to listen for messages than to actually send them. For humans, at least, it's a little expensive generating all that power. There's also the problem of what the message would be, and why you want it sent. It seems that only zealots would send a message."

Tukey addressed a colleague two seats to his right.

"Volker, you're our resident philosopher. Would you answer?"

Volker, who seemed pleased by such an appellation, began his answer, but he was interrupted by Amy.

"Volker, as you know, I'm a journalist. This means two things. Please make your answer understandable, and also say everything for the record. OK? Also, might I get your business card?

Tukey laughed.

"Volker, you're about to become famous, at least for fifteen minutes! Don't let us down!"

Volker, who was a professor at an Austrian university, fished a card from his shirt pocket and proceeded.

"As for speaking for the record, as you say, we've been well schooled on that after the climate change email fiasco. I think the university implanted some chip in my brain when I was napping at my desk to ensure compliance. Oh, was that for the record, too? Whoops."

"Don't worry, that part is already deleted."

"OK. The Message. First, it might not even be intentional. We may have caught a sidelobe from some point-to-point transmission."

"Sidelobe?"

"Directional transmissions are never perfectly directional, at least in our technology. Your main signal is directed where you want it to go, but there are signals off to the sides, sidelobes, that are sent as well. These are never as strong, but they do exist."

"So," said Amy, "We might just be eavesdropping on some alien communications, maybe a weather report; or, worse yet, an alien soap opera – "All My Tentacles."

Everyone laughed, and Volker continued.

"I would hope that my career isn't being wasted on trying to decode some jokes. They're never funny when they need to be explained. That brings us to another possibility. It might just be some data sent from some space probe. The message-senders are still limited by the speed of light, as we are. The message seems to prove that. It's always easier to explore robotically than in person."

"OK," said Amy, "That would be one answer to another question, why send a message at all. If it's just photos of some planets and other data, that would make sense."

"What we're hoping for, however, is a "real" message; that is, an encyclopedia sent by one sentient species to another. In that case, they would try to make it easy to decode. The worth of such a message would be immeasurable , especially if the message senders were more technologically advanced than we are."

"I'm wondering about this technologically advanced angle," said Amy. Do think that their science had a parallel development to our own? Perhaps their medicine is not as advanced; and, anyways, would their medicine even be relevant?"

"We'll just need to wait and see."

At that point, their meals arrived, and everyone was more concerned with eating than talking. Now that she had an entrée into the inner circle, Amy was certain she could conduct some follow-up telephone interviews in later weeks.

The flight from Zurich to New York was scheduled to take slightly more than eight hours. This gave Amy time for some napping, and time for thinking. She used to dread going through US customs with her laptop computer, especially after all the published incidents of journalists having problems. She simply solved that problem by not bringing any computing devices aside from a "dumb" cellphone on international trips.

Her "trick" was to buy an inexpensive tablet computer at her destination for checking email and interacting with the magazine's "cloud servers." They hosted all the applications she needed to take notes, write, edit, and post stories for her editor. All such data was sent encrypted to the servers, where it resided safely without fear of its being impounded or otherwise lost. After "wiping" the tablet computer, she always left it as a "tip" for the housekeeper. She left it in an envelope with a short explanatory note, in several languages, so it wouldn't needlessly be put into lost-and-found; or, worse yet, sent back to her, with shipping charged to her credit card.

Amy didn't even have a telephone number directory on her phone. She memorized the important numbers, and the others were stored on "the cloud," also in encrypted form. If you detained Amy, all you got was Amy, and her publisher would make certain that she wasn't detained for too long.

As she drifted in and out of sleep, she wondered what message she would send. She was a writer, after all, and she could spin a tale as well as anyone. She laughed. What if The Message was the book of a self-published author who couldn't interest Tentacle Press in publication.

The message wouldn't be about her life - Too boring. Something like the history of mankind, warts and all? Then again, perhaps such honesty might not be good. Still, it didn't matter, since the supposed location of The Senders was too far distant for our worrying about a visit. Even if they did visit, it would be so far into the future that we would be different; or, perhaps, not even here.

She must have been tired, since she slept soundly. She was awakened by movement and noise in the cabin as they were landing at New York.

22 ENIGMA

Roos wasn't comfortable around other people. Even minor interruptions would crash his train of thought so thoroughly that it would take him many minutes to get back on track. Hired hands had been arranged to tend his chickens, although Joe DiGiuseppe's choice for the job, the Secretary of Agriculture, was not among them. Still, Roos missed the comfort of his own house and his own bed. It wasn't difficult for Roos to convince the LGM organization that he could effectively telecommute most days, and then venture forth on the occasional business trip when needed. As a perquisite of his new position, he was chauffeured wherever he went, the reasoning being that his time would be better spent thinking in the back seat than driving.

Rick had a small team of software engineers writing pattern analysis software, but his experience with financial analysis convinced him that the software should not be completely automated. When you've honed your software to too accurately find ducks, you're likely to miss out on the fact that there are similar creatures called geese that are far more interesting. Rick didn't want to fall into the duck-goose trap by only looking at The Message through a coded version of human eyes.

Instead, Rick conceived the analysis software as a portal on the data - A way to aid in human deciphering of The Message, as it was now known. Roos approved this concept quite quickly. He had always churned numbers in his head, looking for patterns, and he wasn't going to let a computer second-guess him.

An unlikely contributor to the program was Mary Christoffel, Rick's wife, who had asked Rick the question, "It's radio. Have you listened to The Message?"

"Mary, it's not radio like that. It's just a bunch of numbers being transmitted on radio."

Mary assumed her, "No, I'm not that stupid," stance and explained.

"What I mean is, you're putting graphics all over a computer screen and using your eyes, only. What about using your ears to look for patterns, not just your eyes or those crazy computer programs."

Rick took Mary's suggestion and included audio representations of the data in which patterns could be discerned by frequency changes. He had written software to play parts of The Message as a tune. Not surprisingly, Roos thought that this was a waste of time, but Roos wasn't the only person who was examining The Message, and quite a few others liked this option. Whether it helped in decoding The Message, or not, it did let people "feel" The Message. It was also good for sound bites on the news channels.

This wasn't a static problem. The Message was growing larger day by day, and it had not started to repeat. On one hand, this was good, since the more alien bits the world had, the richer it was, in some abstract sense, but only if The Message could be decoded. NASA, which was in charge of archiving the raw signal from its own radio telescopes and a select few outside its organization, dutifully spliced the signal into one continuous stream and demodulated it into the ones and zeros that fed Rick's database.

As the weeks passed, Rick's activities had devolved into somewhat of a routine. Form time to time, Rick needed to install some analysis programs contributed by The Message community, but his principal task was to ensure that the computer was functional and accessible at all times. There were thousands of registered users, all of whom were vetted by The Message Administration to confirm their qualifications in decoding The Message. Even the smallest of countries took pride in having a team of message experts working on the problem.

Something of such historical importance attracted the attention of more people than just The Message decoding teams. Historians were involved, and they were tasked with the problem of recording an account of the decoding effort. Archivists were working to preserve The Message itself, both in its raw form of digital bits; or, its eventual decoded form. Tukey and his post-doc had been interviewed multiple times. Roos had denied all interviews, claiming that an interview with him would be worthwhile only after The Message was decoded.

The archivists were fast at work on their preservation efforts; so much so, that many of The Message team claimed that they were making too many demands and generally getting in the way. The archivists pointed to the examples of the lost data stores of the space age, including the original recording of the first footprint on the Moon.

The archivists' battles were not just with The Message team. They were also battling among themselves as to the best means of preservation. Digital archives were never very permanent. The media itself degrades over time. Magnetic bits of matter lose their magnetism, and the reflective layers in optical disks oxidize and become less reflective. The usual option is to regularly re-record data onto newer media.

There's the further problem that recording technology changes over time, and common data devices of the past become extinct. A floppy disk saved from just a decade ago would now have no disk drive on which it could be read. The leading contender as an archive format was a high-tech form of microfilm, since magnification would be available to any technological society. In this case, no photo-chemicals would be used, since chemicals always have a way of reacting with their environment and destroying their media.

The digital message would be encoded as long and short marks to indicate the ones and zeros. The marks would be formed from a platinum-nickel alloy on glass, and these marks would be overlaid with a protective glass coating. The alloy had the other desirable property that it was magnetic, which would also aid detection.

Thought was also given to how the archive might be preserved for a subsequent technological generation if our present civilization collapsed, for one reason or another, and the world devolved into a medieval society. Taking its cue from medieval times when ancient knowledge was preserved by monks in monasteries, it was proposed that the archive should be put into the custody of a society of "Message Priests" who would maintain rituals for safeguarding these religious artifacts and induct new priests into their order to preserve the "Message Brotherhood."

Should this happen, Rick imagined that the reception of The Message would evolve into a legend long after our technical civilization had vanished. He presumed that his role in the legend would be that of a bearded scribe to the fairies. It was something to contemplate while resting his fingers between lengthy sessions on the keyboard.

Although Rick was a qualified system administrator as far as scientific circles were concerned, he was never formally instructed. All his tutorial was from actual hands-on experience, and his computer system was supposedly secure by current standards. He kept up with the regular security upgrades, as developed by the development community at large; but, he also realized that his necessarily open system would be a target for malicious hackers.

There was an incessant barrage of inconsequential system attacks, typically ascribed to what were called "script kiddies." These were young, would-be hackers who find some older hacking code on the Internet – scripts – and think they could successfully break into one computer system or another by using it. The development community was always several steps ahead of the kiddies, so these attacks merely suppressed some bandwidth to the outside world and did nothing more.

No one had tried a denial-of-service attack in which a myriad of spurious messages are sent to a computer in an attempt to choke all its Internet connections and deny legitimate users access. Rick knew why. These exploits were usually against smaller companies who could be blackmailed by criminal hackers for a small amount of money to make it stop. These were not very effective against larger companies with a knowledgeable computer staff who knew how to render these ineffective quite quickly. Rick's computer, backed by the US government, could never be blackmailed.

Two real worries were phishing, and a “zero-day” attack. Most phishing schemes are social-engineered methods of getting a legitimate user to divulge his password. They might also involve getting a user to unknowingly download a program that works in the background of his computer to log his keystrokes, including his passwords.

Zero-day attacks utilize some subtle error in the computer system never suspected to exist. These can be quite arcane errors that involve sending a few more characters than were expected in a message, or sending a certain type of message immediately after another. Once these errors are known, they're easy to fix, but on the first day that they're discovered by some hacker, day-zero, they have the potential of crashing a system; or, far worse, compromising a computer system to give a non-authorized user access to do anything he wanted.

Unknown to Rick, the computer literate members of the Lunatic Fringe had been scouring the Internet for possible zero-day vulnerabilities in his computer system. They had begun just a few weeks prior, after Jake, the Prime Lunatic, had given his OK. Their job was was made somewhat easier when details of Rick's system were published in an initial press release about the international LGM committee, newly named the Message Consortium.

Their first step, access to Rick's computer as a permitted user was easy. At the Message Consortium Conference, agents of the Lunatic Fringe had salted the conference hotel with about a hundred memory sticks, all marked with the conference logo on the outside, and carrying a slew of public conference materials, such as the meeting agenda, as their digital content. They hoped that some delegates would takes these at face value and later plug them into their office computers.

Along with the innocuous content was a hidden program that installed itself onto the victim's computer. That program allowed the Lunatic Fringe hackers to capture the user name and password for accounts on Rick's computer. It was all too easy, and it was a technique that spies and criminals use to gain access to corporate computer systems.

Regular users can't do that much. They can create, modify and delete their own files, and they have read-only access to some designated files, notably The Message, itself. Essentially, nothing a regular user might do can cause any harm to the overall computer system. The Lunatic Fringe hackers, however, were not regular users. Armed with a zero-day exploit, they were able to escalate their privileges, and have free access to modify the computer system.

Their plan wasn't to trash the computer. That would have essentially no effect at all, except to shut it down for a day, or so, until it could be recovered from backup files. What they intended to do was to sabotage The Message decoding effort in a subtle way that was less susceptible to detection. They would randomly modify The Message, thereby making decoding impossible.

Modification of The Message, itself, wouldn't work. Such modification would have been easily detected. What they would modify are the copies of the parts of The Message downloaded by each user as they undertook analysis; and, they would do this by modification of the copying program in the computer system. They would simply replace the system program with a modified version they had created. The modifications were designed not to insert random data, but to exchange parts of The Message. Globally, The Message statistics would appear unchanged, but each portion would be scrambled and impossible to decode. Of course, the Fringe was also hoping that The Message was inherently impossible to decode because of its alien nature.

What neither the Fringe nor Rick knew was that the NSA had surreptitiously installed its own monitor into Rick's computer. The installation was necessarily clandestine, since there would have been an uproar of complaint from the Message Consortium, not to mention the many countries who would claim that the US was getting some unfair advantage.

The NSA's monitor wasn't a program running on the system. That would have been detected after a time. While an NSA technician was helping Rick with some necessary hardware upgrades to handle the vastly increased demands for Internet access, he installed a hardware device as an invisible node in the computer network. It was hidden inside one of the many computers in the supercomputer array, and it transmitted a wireless signal to a utility pole outside the computer building. It was essentially undetectable, and it allowed the NSA to monitor everything that happened in the computer system.

Just a few hours after the Fringe exploit began, Joe DiGiuseppe got a call from the NSA monitors advising him what was happening. Joe now had the problem of somehow advising Rick about the security breach without revealing the NSA's involvement. He decided on the “anonymous tip” approach, and he decided to use Roos as an intermediary.

On the telephone, DiGiuseppe, advised Roos that “a government agency” had received an anonymous tip that someone, possibly the Fringe, was planning an unspecified attack on Rick's computer system. He advised Roos that the NSA would be happy to sweep Rick's system to check to see whether anything was amiss; and, if Rick assented, to do some routine checks over the next few weeks. They would do the checks on-site, under Rick's supervision, to ensure that the NSA's involvement didn't go beyond any of the Message Consortium's covenants.

Rick could do nothing else, but assent. It was an easy task for the NSA to “discover” the security breach and assist in recovery. When Rick saw the devious nature of the Fringe hack, he quickly assented to periodic NSA checks, a decision that was affirmed by Roos and confirmed by a vote of the Consortium. The Consortium did make the proviso that the NSA checks would be under Rick's direct supervision. The NSA back door on Rick's computer was still a secret.

This back door wasn't the only NSA precaution. In its usual style, it had a program in place to surreptitiously record all communications between Consortium members. Its claimed purpose, if discovered, was to root out any sabotage from within. As a secondary feature, if there were any successful decoding of any message parts, they would be the first to know.

The Message community was also active on the public Internet, since the text-only requirement for internal messaging on Rick's computer was thought to be too restrictive. Scientists had always believed in the "picture is worth a 1,000 words" dictum, and that's why scientific publications contained so many graphs. When parsing a paper, scientists look at the graphs, first, then they read the associated text if the graphs are interesting.

Several "wiki" sites for registered users who were not necessarily members of The Message team were placed online, and a lot of analyses were posted there. This was seen as a way to both generate public interest, and a means to capture an occasional prescient insight from someone outside The Message community. Fortunately, these web forums were moderated; otherwise, good content would have been lost in a sea of Lunatic ravings.

These wiki sites had further spawned other sites with summaries of topical areas such as advanced signal detection technologies and pattern analysis. The Message was already having an affect on humans, even in its encoded state. Universities had seen an increase in the number of students interested in science, mathematics, and engineering.

One idea that Roos adamantly rejected was an automated release of the raw data stream of The Message. He had nightmare visions of the many web sites claiming to have decoded the content of The Message according to one prophesy or another. He didn't want the true decoding to be diluted by such content. Roos saw himself as not just the administrator of The Message Consortium, but as the guardian of The Message, itself.

23 SETI WHO?

The signal kept coming, and aside from discovering a few number series here and there, there was no handle on decoding. Were these series there as a key to subsequent parts of The Message, or were they there just because they were important? Roos thought that the data blocks adjacent to obviously trigonometric series must contain all the principles of trigonometry, and he set out to see if that message portion could be decoded with that insight.

Any textbook on geometry or trigonometry will always have diagrams, but the mathematics of geometric and trigonometric operations are valid independent of the triangles and other objects it represents. Trying to transmit images would be both wasteful and unnecessary. Analysis seeking to decode those message parts as images was not working.

Early in the project, one team member had a different idea of the purpose of the Message.

"If you tune through the shortwave radio bands, what do you hear? There's a little telemetry, the BBC, Voice of America, Radio Moscow, and a few other international stations. There's still a lot of ship to shore, although the big cruise ships with their huge bandwidth requirements and semi-stable platforms all use satellite links. The smaller ships can't track a satellite on their moving platforms, they don't need the bandwidth, and they can't afford the expense of a satellite link. The other stuff is all religious. At least twenty percent of the broadcast stations are religious. You'll hear Muslim calls to prayer and a whole lot of Christian programs. Everyone wants to convert everyone else, or at least keep the converts on the proper path. I think the signal is religious."

"A lot of others think the same thing. Quite a few cults have sprung up since the discovery of the signal. Naturally, the 'New Age' and UFO followers were the first to climb on board, but a lot of Christian sects are wringing their hands about an LGM theology. For instance, did Christ appear on both worlds? Should he have? We're no longer the center of the Universe, and that makes many people uncomfortable."

"Well, it might just be a cultural thing, sort of the BBC for the colonies. As far as we know, the speed of light is still limiting. The radio signal proves this in a way. If the LGM are an advanced civilization that found a loophole in the speed of light, they wouldn't be transmitting, they would be visiting. The signal is a way for the LGM to keep cultural continuity over the expanse. Something like this would also prevent problems later on. You don't want the colonies to revolt."

If The Message was intended as an advertisement for the presence of a civilization, that left one very big question. Why would a civilization pour so much energy into such an effort? Listening is easy, but broadcasting so many megawatts, perhaps terawatts, was another matter altogether.

Perhaps the civilization was so advanced that their energy stores were essentially limitless. The Message might in fact be the result of a hobby pursued by an individual to allay boredom. It could be presumed that individuals in an advanced civilization would have a huge lifespan, and it might be hard finding things to occupy one's time.

One other strange idea was that broadcasting such a message was actually a way to shed energy. It was a form of thermal management, and the content was just added as an afterthought. If your planet was getting too close to its sun, you could shed energy by broadcasting terawatts of radio waves into deep space. In that case, the Earth should not consider itself to be the lucky recipient, because there were many such transmitters beaming signals in all directions.

The Message might also be a pilot beam directing robotic space craft into the universe. In that case, also, Earth's message would just be one of many. Most of The Message content might contain updated instructions on where the craft should go based on new data from the home world. Roos was hoping that was not the case. Such a message would be far less interesting than a SETI message, and much of it might be impossible to decode for lack of references.

There were a few basic reference points common to Earth and the senders of The Message. Since the laws of physics were the same anywhere in the universe, or at least in the Milky Way Galaxy where both civilizations reside, there were known reference points of length, time, and mass. Although some might argue that one length unit, the wavelength of the signal, was only useful when human-sized objects were involved, there were practical units ranging from the distance between pulsars to the size of an atom, and even smaller. In one strange experiment, it was even possible to distinguish left from right.

This got Roos thinking about how he would describe the Earth and the human species to another civilization. The trivial matters of the structure of our Solar System would be easy, as would the mass and size of the Earth, its sister planets, and their moons. After that, the idea of the Earth being a certain percentage water would be important; and, somehow, the idea that life on Earth evolved from the oceans. The chemical composition of Earth's crust wouldn't be too hard to express, since the atomic numbers in the Periodic Table were an easy reference.

The multitude of plant life, and even the idea of vegetation, itself, would be difficult to convey. The idea of what makes a biological human would be harder, still. There could be no assumption that DNA was the universal building block of life, so the idea of Watson and Crick's 1953 DNA paper would need to be conveyed. Even then, would DNA be understood as making possible so many species? Imparting information as a monologue would prove to be difficult.

Then again, would some SETI species be as interested in humans as they would be in the local space weather? Humans, as a species, are curious, possibly as the result of a survival instinct. Knowledge of the world has its evolutionary advantage, but the warping of this trait into a need for Hollywood gossip and supermarket tabloids might not exist in a sufficiently advanced civilization. A civilization of robots, however they may have come to be, might not see one piece of data being more interesting than another. The average temperature of Earth and the appearance of its human population might be equally important.

Roos was reminded of a humorous article he had once read about some space aliens visiting Earth and deciding that the automobile was the dominant species. If humans were not previously humbled by the idea that the Earth was just a small speck in a huge universe, the discovery of the SETI message and the idea that there are other, perhaps better, civilizations out there should drive that point home.

Humans, however, had always considered themselves to be unique. Aside from their dominant position that caused the extinction of so many species, they were always hostile against any knowledge that denied them a central place in their universe. They first battled against the Copernican solar system that put the Sun, not the Earth, in the center of things.

Even today, scientific evidence is not enough to convince some people in the truth of evolution. To them, it's just some plot by the devil. There could be no denial that there was a signal from somewhere in the universe, but the devil, somehow, must still be involved. Since equilibrium seemed to rule everything in the universe, there might be a fifty-fifty mix of planets ruled by angels and planets ruled by devils. We always presumed our planet to be ruled by angels, but Roos had his doubts.

Once The Message was decoded, should there be a reply? Would the reply be pointless, since the senders of The Message and their civilization would be long gone or vastly changed? There was no need for haste in deciding this, since a few years piled onto thousands wouldn't be important. That was all for the best, since funding for such a project would be impossible to secure in Earth's present state.

There were some who argued that we shouldn't reply. Instead, we should re-transmit The Message to other civilizations farther away in the galaxy, and we should do this even if we were never able to decode it in its entirety. A preamble would be attached with Earth's information. Perhaps The Message was a retransmission already.

Second-guessing The Message content was giving Roos a headache. Each scenario has its consequences, and each of those have their own consequences, almost in an infinite regress. You never know when to apply the brakes.

24 RUSSIAN DOLLS

Not all the action lay in the ones and zeros detected by the network of radio telescopes and fed to Rick's computers. Tukey was still looking at the raw signal, managing a team of technicians and graduate students that tried to improve detection. Tukey and his team had always noticed a very slight variation in the timing of the modulation pulses, but they ignored it as being insignificant. There were so many ways that this could happen, from detection of the pulsar "locking" signal, phase distortion in the electronics, to scintillation caused by interstellar dust, so no one wanted to troubleshoot the problem.

The conference had interrupted this sort of work; and, in the case of The Message, which might be fleeting, every day lost was a tragedy. On the plane trip back from Zurich, Tukey had made a detailed project plan on what needed to be done to research this problem, arrive at a definite cause, and implement a mitigation strategy to improve signal detection.

The advantage of graduate students over technicians, as Tukey freely admitted, was that they were highly skilled labor who responded quickly to every whim of their professor. After a short meeting to brainstorm the problem and assign tasks, a team of four students set out to finally make measurements to pinpoint the origin of the problem so it could be corrected.

So that detection of The Message was not compromised while they were working on essential parts of the signal chain, the work would be done only while there was overlap of detection between stations. Fortunately, there was considerable overlap in the continental United States. Tukey took the precaution of advising the overlap observatories when the work was being done, so they would know that they were, for the time being, the only reliable receiving stations. Tukey made certain he would personally be present while the work was being done, to ensure that there were no problems.

One advantage of their now being a part of the Message Consortium was the influx of money that allowed purchase of better instrumentation. A thorough check of the entire signal chain from the antenna to the data recorder revealed no spurious oscillations or nonlinearities. What they were seeing was actually in the signal. It might be intentional, or an artifact of the sender's transmitter, or caused by something in interstellar space.

Now that it was verified that the timing variation was a real feature of The Message, the problem became much more interesting to Tukey, who had let the students do most of the busy work up to that point. His scientific curiosity kicked in, and he went into the overdrive mode he last experienced as a doctoral student.

One new piece of equipment that they now had was a computerized system that allowed signal analysis in some unusual ways. Since The Message was at the 473 hertz rate set by the lock to the pulsar, they could measure the timing variation from pulse to pulse to build some statistics. Their initial finding was that the variation averaged to zero over a long enough time scale.

That was the point at which Tukey decided to program the analysis system to find the time scale at which the timing variation was maximized. Then, averaging the signal over these chunks, see what was present.

What Tukey found was another message embedded in The Message! At the low data rate of about a bit every three seconds, this message-in-a-message could convey only minimal content. Tukey's theory, which Roos also found believable, was that the slow message was a tutorial on how to decode the faster message. The tragedy was that they had missed it for all this time.

Within a week, all the radio telescopes in The Message network were logging this message in addition to the primary message. Shortly after that, one of the analysts noted that The Message had started to repeat. If the transmission continued long enough, and the inner message had the same period as the main message, they would have captured everything in the secondary message that they had missed.

How long had The Message been transmitting? Millennia? How long would it remain? At this point, most were of the opinion that it must have been transmitting for a long period before it was detected. We couldn't have been lucky enough to be listening at the right place at the right time. The implication was that The Message would be transmitted long into the future.

Discovering this message-in-the-message solved a potential problem. Much conventional radio astronomy had been put on hold, since much of the operating time for many radio telescopes had been assigned to message reception. The conventional radio astronomers were patient, at first; but, as the months passed, they were getting agitated.

If the main message had started to repeat without the detection of the inner message, there would have been a loud chorus of conventional radio astronomers chanting, "OK, guys, you got what you needed. Now, give us our telescopes back!" It was presumed that the inner message had the same period as the main message, so there was ample justification to verify the main message while the inner message didn't repeat.

Detection of the inner message was further confirmation to Roos that The Message had been transmitted for quite some time. The Senders wouldn't presume that the inner message would be found immediately after the main message was detected, so they would patiently repeat.

25 DECODING

The main element of The Message had still not revealed any of its secrets, and Roos was of the opinion that decoding the sub-message was required before decoding the main message. Perhaps the sub-message needed to be read in its entirety before any sense could be made of The Message itself. There were no clues as to the length of this message. Perhaps the sub-message was repeated many times in the period of the main message. We were in elementary school, and we needed to get past the Dick and Jane books before we could understand Shakespeare.

It seemed obvious, also, that The Message was partitioned into a message and a sub-message for a reason. Only those with sufficient technology to discover the sub-message were deemed worthy of understanding the main message. Roos, who was well read in fields other than mathematics, was quick to quote the poet, Alexander Pope, who wrote three centuries ago that, "A little Learning is a dang'rous Thing."

The fundamental problem, at first, was the fact that the transmission rate of the sub-message was extremely slow, so there wasn't much of it to analyze. At its seventeen bits per minute rate, the human equivalent of about two pages of text were being transmitted each day, and what was transmitted wouldn't be text; rather, it would be a contorted series of symbols presumably designed to be understood by an alien species.

One approach to decoding was to imagine how we would design such a message. This was a question bandied about from the earliest days of radio astronomy, and a simple message was actually sent in 1974 towards the star cluster M13, about 25,000 light years from Earth. The message was a binary stream of 1679 ones and zeros, 1679 being the product of the prime numbers, 23 and 73.

This was a graphic message in a 23 columns by 73 rows format. This message from Earth contained binary representations of the numbers, one through ten, the atomic numbers of the five principal elements from which organisms are made, a representation of the DNA molecule, a crude image of a human, a diagram of the Solar System with the Earth highlighted, and an image of the Arecibo telescope, which transmitted the message.

While parsing the low data rate message, Roos noticed runs of zeros much longer in length than expected from statistics. The section these were in was bracketed by two extremely long runs of the binary digit, one. These bit "frames" bracketed 57,599 bits, which was two days worth of data. Curious about the number, Roos found that this was the product of the prime numbers, 239 and 241. These were not just primes, they were "twin primes," members of the set of prime numbers as close together as prime numbers could be.

Roos quickly formed the data into a graphic. The graphic, which had a much higher resolution than the image sent from Earth to M13, had a large, central circle surrounded by a sea of squares, each of which contained a binary pattern, and one triangle with a single dot in its center. The Message had yielded its first communication! Roos immediately organized a team to analyze this image.

It didn't take long to discover that the image squares, fourteen in number, contained the numbers two through fifteen. These four bits of information were about all these squares could possibly contain. The single dot inside the triangle likely indicated that it was the first object. The triangle and squares seemed to be distributed on the image at random, and the numbers weren't in sequence from top to bottom, bottom to top, center to edge, or in any other obvious way.

Looking for clues beyond the image, it was found that the subsequent bits could be decoded as a legend for the image. The numbers zero through fourteen were found in a 64 bit spacing with pairs of other numbers interspersed. It was presumed that the zero entry was associated with the circle in the center. Whatever these other numbers were, they were associated with the pattern in the image.

These numbers were placed in a table where it was found that the entry for the triangle, the first numbered element was an exact power of two. The entry for the circle, "zero" in the table, contained just a single number much larger than any of the other numbers in the table. Now came the hard part - What do all these numbers mean?

The analysis team posed the problem to the entire Message community, and waited. The wait wasn't long. Within a few days a radio astronomer had the answer. The image was not too different from another SETI image sent from Earth. In that case, "sent," didn't mean via radio. Actual objects were sent.

These were engraved plaques attached to the Pioneer spacecraft, the first spacecraft to leave our solar system. These plaques contained some basic information about the builders of these spacecraft, including the location of the Earth. Earth's location was specified by a simple chart that indicated the distance to fourteen pulsars, identified by their periods. These periods were specified in units of a certain electron transition of hydrogen, the most abundant chemical element in the universe.

The radio signal of The Message, of course, was locked to the frequency of a particular pulsar. Assuming that this pulsar was symbolized by the triangle, the radio astronomer reasoned that the entries for the squares were the frequencies of some other pulsars, listed as a ratio to the known pulsar. When he did the calculation, he found close matches to nine known pulsars.

If this was true, then the other numbers must specify distance. The obvious unit was the transmission wavelength of The Message. Using this logic, the numbers seemed to make sense as distances between pulsars. Plugging the locations of these pulsars into a trigonometric calculation gave the presumed location of the planet from which The Message was sent!

As a bonus, the number entry for the circle, the presumed symbol for the planet of the senders of The Message, gave 2.4 years when interpreted as a time. This would be the orbital period of the planet around its sun. Since pulsar periods change with time, the difference between the observed pulsar periods at Earth and the pulsar periods in The Message indicated how long ago The Message was composed. Since radio waves travel at the speed of light, we now had an accurate distance to the planet of the senders of The Message.

Since the senders of The Message now seemed more like real people, they quickly gained the appellation, The Senders. This first communication from The Senders made front page news, and the pulsar-coded image quickly appeared on posters and coffee mugs.

With public interest so high, The Message Consortium knew that funding would not be a problem. The public, however, wanted more. Was this easy decoding just a fluke, or would the rest of the tutorial be just as easy? Then, beyond the tutorial, what about the main message? The Message was exciting, but these unknowns were what kept Roos awake at night.

The obvious next step was a computer program that automatically sifted through The Message looking for similar graphics. Not much of the sub-message had been captured at that point, so it wasn't surprising that it contained no other graphics. Surprisingly, the main message contained no graphics at all, at least no graphics specified by the rather broad assumptions of the computer analysis.

The absence of graphics in the main message confirmed to Roos that the sub-message tutorial was critical to the decoding process. In one sense, this was frustrating, since progress was locked to the slow data rate of the sub-message. Roos remembered the serials in the Sunday comic pages of his youth, and how he hated waiting a full week for resolution of some plot point. It must have been the same for the previous generation, who often waited much longer for the next episode of Flash Gordon or Buck Rogers to appear at the local movie house.

There was another reason why Roos yearned for the next graphic. He was interested in how often they would appear, and receipt of a second graphic would give him a basic idea of what that might be. The more graphics received, the better the estimate he could make. For now, he would just need to be patient.

Mathematicians are more likely to labor in solitude than collaborate on problems, but most of science revolved around teamwork, and some interesting discoveries in science had been made by interdisciplinary teams. Scientific specialties had evolved for the important reason that there was so much prior art a practitioner needed to learn before making a novel contribution. Specialists, however, were too focused on their small segment of the universe to see the forest, and not just the trees.

Although himself a mathematician, Roos saw how bringing people of different backgrounds together stimulated ideas. That's why the Message Consortium had a few philosophers and sociologists on its staff to complement the linguists, mathematicians, and computer scientists. It was a way to generate new perspectives on The Message, and potential clues to its decoding.

Taking the lead from the Internet giants, Roos ensured that the headquarters coffee room was spacious, comfortable, and stocked with more than just coffee. He wanted a setting where his staff would enjoy spending time away from the work that they perceived to be more important than a coffee break. He wanted them to sit for a while and chat about their work with staff from other specialties.

Roos, himself, made certain that he, too, took frequent breaks so that he might overhear something of interest and perhaps interject his own thoughts. Although he was the supposedly busy Director, he kept the door to his office open, and he encouraged occasional walk-ins. He knew that the staff wouldn't abuse this privilege.

26 AMY AND JAKE

Frequent mention of The Senders of The Message in the press was beginning to grate on Jake, the Prime Lunatic, so he was happy when he received an unexpected telephone call from a journalist who wanted to get his side of the story. She must be good. His home number was unpublished, but she had tracked him down to his work location, and she telephoned him there.

Of course, all this was a further sign for caution, since she might be a government agent. She had requested a telephone interview, but Jake asked for a face-to-face. He looked her up on the Internet and decided that Amy was genuine. He was happy that her photograph appeared in multiple places, and in articles published in prior years. He just needed to make sure that the "Amy" who arrived at his doorstep was really the journalist.

Amy's cross-continental trip was uneventful. At her editor's "suggestion," she traveled on a Sunday to get a lower airfare. At least her magazine wasn't stingy when it came to hotel bookings, since they had a contract with a major chain. They might also have qualms about sending their female journalists to cheaper accommodations in the more dangerous parts of towns, but she thought this was motivated by a fear of lawsuits rather than some actual concern.

Although she had suggested a dinner meeting at her hotel, Jake, the Prime Lunatic, said he would prefer that she see him at his apartment after work. Jake disliked most public places, and he figured he would have more control of the situation at his apartment. They made an appointment for 7:00 PM, which would give Amy enough time for a good dinner and Jake a chance to unwind from his day's work. Amy knew that distracted people gave short, and bad, interviews.

Her rental car's GPS got her to Jake's apartment without any trouble. She found herself in a low-budget, three tier, apartment complex of the usual layout. Each apartment had a balcony that opened onto the scenic delights of the parking lot. Although some of the apartment dwellers were on their balconies, barbecuing on small propane-fired grills, she thought that the true purpose of the balconies was as watch stations for their automobiles. This was not a nice part of town.

Amy buzzed Jake's apartment from the entry way kiosk, and her buzz was returned by the buzz of the magnetic door lock. She walked to the second floor to find Jake waiting for her outside his door.

"Hi," said Jake.

Amy extended her hand.

"Hi, I'm Amy Pomronger. California is always a nice respite from New York, although the traffic seems about the same."

"Yeah. It's even harder when you need to travel during rush hour to get to work. Luckily, my employer doesn't worry about when I arrive. He just wants his ten hours each day."

Amy laughed, but this didn't appear to be a joke.

"So, they work engineers pretty hard out here?"

"The problem, as I see it..." Jake paused to compose a sentence that wouldn't offend, "...Is that there's too much competition from visiting engineers, and offshore engineers. They're pulling us down."

Amy decided not to comment, and she was invited into Jake's apartment. It was what she expected for a bachelor technology fan. There was a computer in the small space likely called a living room, a wide screen television, and what looked like some gaming systems. Aside from a few chairs and some snack tables, there was very little furniture. She could imagine what the bedroom was like, but she didn't want that image to linger in her mind.

"Have a seat."

Jake directed her to what was likely the best seat in the house, and he pulled up another chair opposite to her.

"Oh, can I offer you a drink, or anything."

Amy was a little thirsty, but she was hesitant to accept anything in a Jake-washed cup.

"If you have bottled water, cold, or not, that would be nice."

"Sorry no bottled water, just tap."

"My editor told me not to drink the California water. He was afraid it would entice me to stay here; so, no thanks."

Now Jake laughed.

"So," Amy continued, "The best way to do this is to let you give me your history, the history of The Lunatics, as they are called, and your ideas about the Message. Then, if I have any questions, I'll ask them."

Amy continued.

"First, do you talk with the press that much? Could I call this an exclusive interview?"

Jake pushed back on his chair and paused for a few seconds.

"When all this started, I got a few emails from some journalists. I answered these, and then I found that my words had been twisted. What they wrote made me look ridiculous. I'm an educated man!"

"I stopped answering emails, so I haven't communicated with the press in about a year. Then I hear from you, by telephone. You're from a respectable magazine, and you were interested enough to interview me face-to-face. I really want to get our point across - *Our,* meaning the Lunatics - but I don't want it distorted. And, this isn't just my viewpoint, it's the viewpoint of everyone in my organization."

Amy was starting to have some sympathy for the guy. Her article wouldn't read like a supermarket tabloid, but she still needed to write the truth. Jake would get fair treatment, but she wouldn't hold back any punches.

"Good. I'll call it an exclusive interview. Oh... and I'd like to record this, so I don't get anything wrong."

"Recording is OK. I'm going to record this myself," said Jake.

"Whoa," thought Amy. "I knew this guy was a little paranoid, but this tops it."

"Yeah, that's fine," said Amy. "Turnabout is always fair."

Jake gave a summary of his life, which might have been the life of about a million young men in the country. He did well in high school, but never excelled. He ended up in a two-year college for lack of money, worked while he attended school, then was able to spend his last college years at a state college, studying engineering.

He was an electrical engineer, a fairly good computer programmer, and he had been working at the same large corporation for more than a decade. He had seen a lot of changes in his profession during that time, none of them very good. Amy could see how his resentment had built up over the years. She could see parallels in her own profession.

Being computer-literate, Jake spent nearly all of his non-working time on the Internet message boards. He had spent so much time on the technical message boards, in fact, that he had become well regarded. That was how he floated the idea of "The Lunatics." The Lunatics had followers on many message boards. The "membership," if that was a valid term, was tens of thousands, but Jake communicated privately with a select group, and his opinions filtered down through the ranks. Most in this inner circle were known only by their alias, and it was a rare occurrence if any of them ever met face-to-face.

Some had tried to identify the Lunatics with one political movement, or another, but their beliefs were too eclectic to be pigeonholed that easily. Their opinion of The Message seemed to be the same as "The Message is evil" stance of various religious organizations, but they didn't think that The Message was evil in that sense. They believed that The Message was an unwelcome interference with human civilization and that humans should be masters of their own destiny.

With Jake's background material and her questioning, Amy had more than enough material for her article.

"Just one more thing, Jake, since the article is about you as well as The Lunatics, I'd like a photograph. Perhaps you could pose in front of your computer, or something like that."

"I can do better than that."

Jake reached back to his desk and retrieved a manila folder.

"I had these made when I thought the press would take us more seriously."

Jake passed the folder to Amy, and when she opened it, she saw a portrait photo of Jake. In the photo, Jake looked more like someone running for Congress. He was dressed nicely in a suit, and he looked not at all like the rag-tag engineer she saw before her. Jake laughed.

"That's my wedding and funeral suit. I don't get to wear it much."

Jake escorted Amy to the front door of the building and waved as she pulled away in her rental car. Jake was nice enough, but Amy left the interview worried about how many people like Jake might be in the world; namely, tech-savvy people who could do serious harm to advance their cause.

A cantankerous old neighbor squirting his hose at the school kids who cut across his lawn is one thing; but technical people had dangerous skills, and Amy was starting to realize how society's alienation of this class of people was bad policy. The outsourcing of technical jobs in the quest for maximum profit had led to unemployed engineers, and marginally employed engineers like Jake. Jake had found solace on the Internet, where he was espousing some fringe ideas, but doing no more than getting eye strain from looking at a monitor all day. How many more were loners who sat at home at night plotting revenge?

Amy had no idea how to make a bomb. Jake and his peers had never taken a course entitled, *Bomb-Making 101*. However, knowledge from their chemistry, physics, and engineering courses, when combined, was the same thing. Although science and technology were male-dominated professions, Amy was certain that a female engineer could make a gun out of a tampon tube if she wanted to.

As Amy hashed-out the draft of her article about Jake and the Lunatics in her head, she thought about incorporating some of this, but she decided against it. Her article was about The Message. That's why people had come to read it. "Let's leave the political stuff to the other rags," she thought, as she worked her way though the lighter evening traffic to her hotel.

Amy replayed the interview in her hotel room, annotating and refining some points in her handwritten notes. Some of her colleagues were all electronic, taking their interview notes on laptops or tablet computers, but Amy wasn't that adept with a keyboard. The handwritten notes were faster for her; and, although they added another step to the article writing process, this added step did lend itself to further reflection on the topic.

Jake's Internet cohort was exclusively male. It wasn't a consequence of their hanging a "He-Man, Girl-Haters Club" banner on their message boards. It just happened. Perhaps this electronic version of the rough corner of the schoolyard appeared too dangerous to women. More likely, the women had more interesting real-world things they were doing.

Another interesting aspect of the interview was the idea that Jake and his colleague engineers had been developing the technology that eventually puts them out of a job. A decade ago, computers had decimated the huge corp of draftsmen who had worked for such companies. Now, one person could accomplish the work of ten. The Internet had enabled companies to utilize lower wage engineers anywhere in the world. It didn't matter whether they were just half as qualified as the local engineers when they worked for a quarter of the wage.

The same thing had happened in Amy's profession, where she often had to compete with free talent in the form of unpaid interns who thought the experience would be good leverage to a paid position. Amy laughed. If they really knew the tenuous status of the position they were trying so hard to get, would they still be willing to work for no pay to get it?

Amy prepacked a few item in her suitcase and headed for bed. She had an early morning flight, and the time change from west to east was sure to exact its toll when she returned to New York. Still, she wasn't able to sleep. She never slept well on travel, and this trip had given her a lot to think about. Jake and his cohort might be wrong about The Message, but they were right about one thing. The Message would change human history. How, exactly, was yet to be determined.

Having a liberal arts education allowed Amy to put The Message into a different context than Jake's engineers. She saw a parallel in reception of The Message to Plato's allegory of the cave. Humans have been staring into space since cave times, first just seeing a few stars, then cataloging and analyzing many stars with great precision.

To Earth men, the universe had been just stars. Now, the universe wasn't just stars. It was intelligent life, and all those stars were just there for reading light and sustenance.

Amy had started to relax and drift into sleep with the warm thought that we're not alone in the universe. Then again, it might just be that misery loves company.

27 UNDERSTANDING

With The Senders of The Message getting frequent mention, the undercurrent of debate about The Message quickly changed into a torrent. As expected, the "Message from Satan" and the "Message from God" corp became much more active. On a more reasoned front, philosophers and ethicists had something newer and more pertinent to discuss than transhumanism.

Off to the side, but still a potential threat, were the Lunatics. Although the attack on The Message computer wasn't publicized, those in the know had always thought that the Lunatics were the culprits. The enhanced NSA security on Rick's computer assured that something like that wouldn't happen again, but it was assumed that the Lunatics would continue their campaign to stop decoding The Message.

Since the Lunatics might even resort to physical force, such as planting a bomb at the computer, the site security was upgraded. Even Rick was required to wear his badge at all times, and the badge served as the electronic key to two layers of access doors to the supercomputer.

Even the main message was beginning to reveal its secrets. As expected, most of the content was mathematical; and, when it was accessed starting at the proper point in The Message, it began with number theory and simple arithmetic. The trigonometric functions, since they could be expressed as series expansions, came next.

A rehash of known mathematics wasn't what Roos was hoping for. Since The Message had started to repeat and its length was known, Roos realized that its content was too limited for anything of substance. If Fermat's Last Theorem had not already been discovered after centuries of struggle, it would have been interesting to find its proof in the Message. That proof, however, was a huge conglomerate of ideas from several areas of mathematics, and it could never have been packaged by The Senders as part of The Message.

Some people in the Message Consortium were starting to wonder whether the importance of The Message had been overblown. Sure, the initial result of pinpointing The Sender's location was interesting, but perhaps the real importance of The Message was just the fact that we're not alone in the universe. A cure for cancer, world peace, or anything of the sort, would not be found in The Message.

Still, it was their job to decode The Message, so they continued in their work. The decoding of the pulsar image sent Roos on a quest to find similar images in the low data rate message. His idea was that all such images would adapt the 239 by 241 format of that first image, perhaps with ancillary information following in the data stream. The problem was that they had discovered the low data rate message long after reception of the main message, and there wasn't that much of it archived.

The computer analysts had written software to statistically flag message areas in both the low data rate and main message that were likely to contain images. For the low data rate message, they constrained their search to two-level, black and white, images like the pulsar image. The main message, however, had enough of a bit rate to give some photographic quality images in shades of gray. They could only hope for an actual portrait of a Sender. Perhaps they wouldn't even recognize it as such.

Humans had seen many television series and movies with space aliens, but all these were of the same type. What most people expected in a space alien was a humanoid with green skin, perhaps with a tail, or a second set of arms. Although having the skill and dexterity to master radio would pose some boundary conditions on The Senders, the SETI community expected something far different from their space aliens.

If natural selection had taught us anything, it's that organisms are optimized to their environment. Creatures in a high gravity environment would likely be short and stout. Creatures in a rarefied atmosphere would need larger lungs to keep a reasonably high metabolic rate. In this case, reasonably high would mean fast enough to design, build, and operate a radio telescope. Low light would mean big eyes. The list could go on and on, but we knew nothing about The Sender's planet at this point, other than its orbital period of 2.4 years around its sun.

After a few more weeks of reception of the low data rate signal, the software detected another 239 by 241 image, and within a few minutes everyone in the Message Consortium was viewing it on their monitors. This image could be decoded merely by inspection.

The image was a plan view of a solar system, sun in the center, and six planets. The sun and planets were depicted as circles, or what approximated circles at the image resolution. One planet, the second from the sun, was depicted as a filled circle, and all the other circles were blank. The filled circle, presumably, was The Sender's planet. There was also a triangle with a single dot in the center outside the system. The Message pulsar was again being used as a reference. The other symbols were tagged with the numbers from two through eight.

As expected, the image "legend" appeared immediately after the image in the data stream, and the numbers signified mass, as referenced to mass of the pulsar, and distance, as referenced to the distance between the Sender's sun and the pulsar. Pulsars are known to have a mass about fifty percent larger than the mass of our Sun, so a calculation placed the mass of the Sender's sun just slightly smaller than that of our Sun. Not surprisingly, if their sun was similar to our own, the distance from the Sender's planet to its sun put their planet in the habitable zone for life as we know it. The Sender's planet was slightly larger than Earth.

Now that public interest had shifted from The Message to The Senders, Amy's editor assigned her to write a Sender's-oriented article that would be both understandable and interesting to a typical non-scientist reader. Amy's terse reply was, "I've always wanted to write science fiction."

To inject some semblance of science into the article, Amy and an illustrator took the relatively short automobile trip to interview an astrobiologist at Cornell University. At least on paper it looked short, but there was some major delay at the George Washington Bridge, so they were late for a planned restaurant lunch and ended up at the university cafeteria. The professor didn't mind, since he ate there regularly, and Amy picked up the tab.

They retreated to the professor's office in Clark Hall, and Amy began.

"So, Professor..."

"Just call me Carl"

"OK, Carl. The article, of course will cite your full name and credentials. Naturally, I find the credentials to be quite impressive."

Carl laughed.

"In academia, it's more a matter of age than anything else. See the gray hair?"

Carl stroked his hair, and both Amy and her illustrator laughed.

"Jack, here, is one of our best illustrators. Jack will be sketching some conceptual drawings of The Senders as we talk. There are so many things that The Senders could be, but it's important to filter-out things that they couldn't possibly be. It's beyond me how something like this could be done, but we've called on an expert."

"I'll try not to embarrass myself and my SETI colleagues."

The interview was two hours long, and Jack had generated quite a few sketches. Amy drove on the way back, but she was constantly examining Jack's sketches as she drove. As someone so famously said, "The road's straight, the car's going straight, so what's the problem?"

"So, those aren't antennas, they're horns?"

"Yeah, that's what the man said. Please don't give them antennas, but it's OK to give them horns. Life is competition, for one thing or another, but usually a mate, so vestigial horns are OK."

"Then there's that other thing he said about The Senders being asexual - No male or female."

"Don't worry, I'm sure Andrew wouldn't allow depiction of space alien sex, or space alien reproductive parts, anyways."

Amy dredged up some arcane knowledge from her reading.

"Edgar Rice Burroughs had the interesting idea that Martian women laid eggs."

"One thing he said was that hands were essential, and I should show them handling some tech object. Multiple arms were OK, as long as they seemed functional. So were multiple legs, although the planet's size probably didn't need more than four. Tails were optional."

"What about the more exotic lifeforms," said Amy. "He said that they might be amphibian, although a pure sea world was likely out the question since you can't burn coal and stage an industrial revolution underwater. But he liked your insect."

"Yeah. He said you should call them exoskeletons. He liked the sketch of the articulated bug with four feet on the ground and four hands above the joint. He said the bug should be pushing some buttons; and antennas - The bug type - were OK on those. He said that the fuzzy outline with the big question mark was probably the most informative sketch, since we really know nothing about The Senders.

A light rain had started, so it was going to be a longer trip back. Amy didn't care. She had plenty of material for an article, and her time behind the wheel gave her time to run some phrases through her head. The article would practically write itself, although Jack still had a bit of work ahead of him translating those sketches into works of art; and, more importantly, works of art vetted and approved by their editor, Andrew.

What Carl had emphasized was the human tendency to anthropomorphize everything from teddy bears to television space aliens. The human body plan seems ideal to us, but we're definitely prejudiced. What's more important is function; namely, all those things that an organism would need to survive in its world, and, in the specific case of The Senders, to build a radio telescope.

The only requirement, it seems, is a brain. After that, it's your brand of evolution's choice as to how you would eat, breathe, touch, and move. The up-down motion of legs against the pull of gravity wastes energy, so why not slither like a snake? Your whole body needn't slither, just enough of it to make you move.

In thinking about how bizarre these space aliens might be, and how the world was captivated by their message, Amy wondered why such a minor thing like skin color could affect one person's view of another. Maybe The Message would have a beneficial affect on the world independent of its contents. Once again, that was a bit of philosophy better left out of her article.

The most interesting topic, the sex life of The Senders, was brought about by Jack's quip about supermarket tabloid stories of space aliens raping Earth women. Carl gave them a long discourse about why he thought that The Senders were sexed. Although organisms could reproduce by budding, or fission, just as Earth bacteria do, there was a good argument in favor of Sender sex.

Sex involves competition for a mate. The image of the most virile caveman getting the prettiest cave girl had persisted into the present day with the trophy wife. In each case, the women are attracted to the men who can provide the best lifestyle. Secondarily, such competition would have given us our technological civilization. Perhaps, without competition, we would all be grazing on the open fields, and so would The Senders.

The swish-swish of the windshield wipers had lulled Jack to sleep, and Amy redoubled her effort to pay strict attention to the traffic. They would be home in a little more than an hour, and she could be attentive for that long. After that, she would welcome her pillow and a long sleep. With the great background material she and Jack were bringing back with them, Andrew would likely forgive a late entrance the next morning.

28 SABOTAGE

The Lunatic Fringe was quite surprised when its computer hacking attempt to sabotage message decoding was so easily derailed. Jake, knowing the skills of his computer hackers, had believed that the attack would be a great success, but he had a worse problem. His supporters were beginning to lose hope. How could they, a rag-tag group of zealots, derail the efforts of an international consortium of governments.

Of course, one lesson learned from the Vietnam war, subsequent armed conflicts, and computer hackers, is that a small offensive force can defeat a large defensive force, simply because the countermeasures for defense are considerably more expensive than those required for an effective attack. Also, those expensive countermeasures, if not well planned, are not a hundred percent effective.

The Lunatic Fringe, itself, was international, so it was time to rally the troops in other countries. The Australians, in particular, were quite concerned about the possible affects of The Message on their culture. They reveled in their isolation from the world, and they were not about to let an alien world destroy what they held dear.

A nostalgic mythology had developed around Australia's use as a penal colony in past centuries, and many proudly preserved a distrust of authority. The Fringe Australians, bedecked in their Akubra hats, wide belts, and huge brass belt buckles, fancied themselves as descendants of the rough-and-tumble characters of their past. They were not timid about staging a direct assault on the reception stations.

The Canberra Deep Space Communications Complex at Tidbinbilla, Australia, was an early recruit to the LGM monitoring network. It patched a major gap in continuous coverage as the signal swept around the earth every 23 hours, 56 minutes, and 4.1 seconds, or one sidereal day. Canberra was the sole recipient of the signal for a little more than four hours each day.

Canberra certainly wasn't part of the Australian outback, but it was happily free of most of the interfering signals that plagued most of the populated world. That's what made the event that much more surprising. Displays started flashing warnings that there was interference on the Message frequency. The interference was so strong that the Message was being jammed. The problem was at first thought to be a front-end amplifier breaking into oscillation for some reason. Swapping in another amplifier gave the same result, so it was time to look for external sources.

Canberra had a survey vehicle, a van equipped with radio receivers, directional antennas and spectrum analyzers. It was occasionally used to locate radiating television cables and faulty transmitters. A team was quickly dispatched to locate the interfering signal.

That's when things got interesting. The directional antennas located about half a dozen intense sources, all on The Message frequency, and a spectral scan showed that the signals weren't unintentional harmonics of a standard signal. These were transmitters set to purposely transmit on The Message frequency! Not only that, but one of the transmitters was moving. An obvious first target was the moving transmitter, since it was presumed to be in a vehicle. If they could catch the vehicle, or its license plate, the driver could be interrogated.

As they tracked the moving source, it first appeared to be an off-road vehicle, since it wasn't traveling on roadways. The SUV couldn't navigate the terrain, so they could only hope that the vehicle would soon return to a road. Soon, the survey team realized that this vehicle was moving too far away from them to allow interception, so it was decided to abandon this first pursuit.

The next effort was to locate one of the stationary transmitters to discover what they were up against. Reasoning that the strongest of these must be the closest, the team headed in that direction. After a ten minute's drive, they located the approximate area and fanned-out with portable receivers. A shout from one of the team brought them to a scrub brush with a hand-painted sign with a short message relating to the legitimacy of their birth. At the base of the bush was a small box with an antenna.

As one of the team went to disable it, another told him to stop. First, they should take photographs as evidence of what was obviously a criminal attempt to disrupt The Message signal. Second, the box might be booby-trapped.

No one ventures into the Australian outback without a rifle; so, after photographs were taken, the team stood at a distance while one of the members fired off a round. It missed, much to the derision of the other team members, but the second shot hit its mark. There was no explosion, but the transmitter was disabled. They marked the location for a subsequent police follow-up.

As the night wore on, the other stationary transmitters were found and disabled. Since all the transmitters were identically constructed, and the previous ones had not been booby-trapped, the team leader decided to benignly disable one and bring it back to the station for disassembly and inspection. The moving transmitter was still a mystery; but The Message window for Canberra had passed, other stations were receiving the signal, and there were still about twenty hours before such interference would again be a problem.

Examination of the transmitter back at the station showed a very inexpensive construction with battery power enough to last just six hours. It contained a simple timer to activate the transmitter at a set time. The battery lifetime meant that the transmitters were designed for just a single jamming session. The inexpensive construction meant that they could be deployed regularly to interrupt the signal at each passage. More worrisome was the fact that the timers could be set for days in advance, so there might already be hundreds of them, hidden in the underbrush. Canberra may have been a test for a jamming effort of all message stations. This was serious business!

One team member decided that in matters of electronic forensics, his knowledge easily trumped any police experience. He had a possible way of tracking the perpetrators. He wrote up a parts list of the components in the captured transmitter, suspecting that certain combinations of parts on a single order would be rare. When the police arrived, he gave them the parts list, along with a list of the usual hobby electronics sources.

With the police investigation underway, the next Canberra task was to determine the nature of the moving transmitter. The electronics in the van had recorded its signal strength and bearing during the attack. This allowed the computer people to estimate its trajectory. The signal seemed to move at a constant, rapid pace that was independent of the terrain. It just took a little thought to decide that the transmitter was on an unmanned aerial vehicle, possibly as simple as a radio-controlled hobby aircraft.

Back in the US, Jake, the Prime Lunatic, had received an encrypted email message about the successful Australian attack. Finally, they were making progress towards stopping reception of The Message! Pending a successful outcome of the Australian test, there were plans to implement similar attacks at other telescope locations. Because of the remote locations and semi-totalitarian governments for many, they couldn't hit every one, but they could easily hit a majority. Radio astronomers, like everyone else, didn't care to be too far away from civilization.

People regularly devoted their time to the cause, as did the original computer hackers, but the jamming transmitters required an outlay of cash. Since these were disposable, they needed to be replaced on a daily basis. The Lunatic engineers had brought the price of the circuitry, less battery, down to about forty dollars each. Since the antenna could be just a short length of wire, and the enclosure could be as simple as a coffee can, the total cost was about fifty dollars. Not much, but it was still $5,000 per hundred.

There were jokes about outsourcing the job to China for a steep discount, but a paper trail would lead back to someone, and customs would be involved. As it was, they needed to fabricate their own copper circuit boards, since outsourcing even those would be too risky. It was a labor intensive operation, but many were willing to donate their time to the Lunatic cause. They were also smart enough to split-up the component order. Everyone had seen enough crime shows on television to get an idea of how the police operate. It's also nice that the cops in real life aren't as smart as the ones on television.

As for funding for the electronic components, the computer hackers posed a plan to do what hackers usually do; namely, steal money on the Internet. Jake vetoed that idea. The Lunatic Fringe was not a criminal operation. In his mind, they were there to do good for a world that didn't see the inherent evil in The Message. He would just pass the hat, as needed. In the short term, he would dip into his own savings as required.

The Canberra team realized that a single van wasn't enough to hunt down multiple transmitters. They also realized that just hunting and killing the transmitters wasn't the best solution, since too much of The Message would be lost each night. They decided that the best option would be to talk the police into setting up road blocks within a ten mile radius of the telescope to catch at least one of the perpetrators, and this would lead to the arrest of the entire group.

After another day of jamming, the police reported back that the combinations of parts they had received didn't get any matches at the regional component distributors. The perpetrators had apparently realized this evidence trail, and they had taken the precaution to order things separately. The police asked for further advice.

One particular component, an ultra-high frequency transistor, stood out as being an essential component. An order for a few of these, or thousands, would not raise a red flag, but an order for a hundred might. The police were asked to get customer information on that basis.

The Canberra incident reminded Roos why he had retired to his chicken farm. If the slower sub-message never repeated, whatever data were lost were lost forever. Or, as the newspaper headlines stated, "The Cure for Cancer, Lost Forever?" Roos never thought that something like a cancer cure would ever be in The Message. He doubted that alien biology would mimic Earth's so closely, and he was fairly certain that The Message couldn't contain something that detailed. Perhaps The Message might summarize a cure for *their* cancer, but not our own. Also, as physicians pointed out, cancer wasn't just one thing, it was many. No single silver bullet would cure the lot. Still, information lost was information lost.

Roos had immediately chartered a team to determine whether a filter or other electronic or computer trickery could recover The Message in the presence of interference. He was counting on the fact that The Message was frequency modulated, and the jamming at this point was a fixed frequency signal. Even if the jammers advanced to the point that they, too, were frequency-modulated, their signals might be predictably filtered. It was a slim chance, but there was nothing better he could do.

One thing that Roos recommended, immediately, was a continuous, high quality recording of an omni signal at each telescope. He knew that just subtracting the omni signal from The Message would not give an interference-free signal. At the Message wavelength, any small change in receiver position gives a different signal, especially when multiple sources are concerned. Today's technology couldn't reconstruct a clean Message signal with the help of such a recording. Future technology, however, might easily do this, and they would be remiss if they didn't at least capture the raw interference signal for posterity.

29 TRACED

The police discovered a recent order for fifty transistors of the type identified by the Canberra team. These were purchased on a prepaid credit card, and the shipping address was found to be a vacant house in a suburb of Wollongong, quite a distance from the Canberra telescope. The police asked the electronics distributor to monitor subsequent orders for that part.

The successful jamming attempts had emboldened the Fringe team to continue production of the transmitters, so additional parts were needed. They placed further orders, being careful to use prepaid credit cards, and they changed the shipping address to another vacant building. The police, however, got the new address from the parts distributor, and they monitored the building. Just a few days later, they had one of the Fringe in custody.

The suspect was not just a "mule" who was sent to get the box, he was one of the team who built the jamming transmitters. Like his fellows, he was in his late twenties, half disgusted with his life, and looking for adventure. He and his team had easily fallen under the Prime Lunatic's spell, since he was older and wiser than they, and he was a leader. Everyone else would just rant, but Jake, the Prime Lunatic, would plot.

The first rule about talking to the police is to not talk to the police. The captured Fringe member was saying nothing, and he realized that a simple electronics parts order, paid by a legitimate and not stolen credit card, was evidence of nothing. Even if this warranted a search of his apartment, the police would find nothing. The Fringe, if anything, was tech-savvy, and they considered themselves to be quite a bit smarter than any policeman.

In the present age, interrogation wasn't always necessary. The police, knowing the identity of their detainee, had an easy job of piecing together his social network. Within a few days, they had more of the jamming team in custody, and other evidence. Things were looking bad for the Fringe.

Jake was quickly informed about the initial team member's capture, and the subsequent arrest of the other team members, through an encrypted message. He knew that in the short term, there would be no evidence linking the Canberra jamming to him. The long term, however, was problematic, not just for his own freedom, but for the cause itself.

Jake spent the next day deciding what his next step should be. He couldn't just run, abandoning his friends and the cause they had worked towards for so long. He decided it was time for bold action. He summoned his courage, packed his SUV, and set out for the cabin.

Uncharacteristically, Jake obeyed the speed limit on the interstates on his route, but he tried to travel on parallel local roads whenever feasible. He needed to keep as low a profile as possible. Even if an all-points bulletin hadn't yet been issued for him, any speeding ticket would mark his location and get the cops thinking. It would be difficult to trace the cabin to him, since his ownership was likely not in any electronic database. Small, rural governments had neither the time nor the resources to convert their records into electronic form, but there were still written records.

Jake's trip was uninterrupted, except for a quick stop at a self-service gas station, where he paid cash. He hated the fact that his aged SUV burned fuel at such a high rate, but the extra room was a help. On this trip, he brought nearly everything edible from his apartment, along with much of his clothing, and all papers and electronic documents relating to the Fringe. He was effectively abandoning his apartment. That's the first place where the police would look for him and evidence of the Fringe activities. He would burn most of that at the cabin, collect the ashes, and burn those again, mixed with some charcoal. Any job worth doing is worth doing right.

It was hard even for Jake to find the entrance of the access road to the cabin. Fortunately, he had plugged its coordinates into his GPS unit on the last trip. He was happy that the cabin was so hard to find. If troubled times were to come, isolation is the best defense. The road, itself, was hard to navigate. Lack of traffic had ceded the ground back to nature. It was just the lack of large trees along the path that defined where he should drive.

The cabin looked as it did the last time he had been there, more than a year ago. He had planned that trip to be a vacation, but a leaky roof demanded his attention for the first two days. Repairs are difficult when the nearest lumber yard is a thirty minute's drive. He fixed the roof, handled a few other repairs, and he hoped that the cabin would survive the next winter without any other problems.

Jake's first action after arrival was to unlocked the cabinet and look inside. The gun was there, a forty-five caliber Glock pistol. He removed the gun, which was cold to the touch, and examined it. It was at least a decade old, in very good condition, but it should still be cleaned and oiled before use. The cleaning kit and oil were there in the cabinet, as were about a hundred rounds of ammunition.

Jake checked the ammo clip. It was empty, as it should be. He took the empty gun, left the ammunition inside the cabinet and locked it. He wasn't a criminal. The gun was to intimidate, not to use. He stuffed the gun into his oversized coat pocket.

One thing that his hacker pals had done before the arrests was to track down the home and business addresses of the principal Message characters. Most were foreign, and the US contingent was spread all over the continent. Roos was within a day's ride, but couldn't be expected to be at home; or, alone at home, when he was there. Two minor characters were the supercomputer guys, who were quite close. Jake had no intention of entering the supercomputer building. There were too many people, and it was in the middle of the city. Hostage-taking like that was for losers, and Jake wasn't a loser.

What, exactly, was his plan? He would eventually be captured. That was a certainty. His only logical course of action was to get his message out to the world. Maybe it wasn't too late late for the people, themselves, to stop their governments from infecting the world with The Message.

Jake drove off in his SUV, still undecided whom his target would be. He needed to kidnap someone in the Message circle to make his point. And, he would need to parlay that kidnapping into a major press event. Women, of course, are a far easier target than men, and they would get better press coverage.

Parking his van after a drive-through at a fast food place, Jake studied the addresses and the thumbnail biographies that the Fringe had assembled. The Christoffel house was close to an interstate on-ramp; and, from the biography, its looked as if the spouse was a housewife who was likely to be home. Jake decided that would be his first target. If no one was home, he would try the Rasmussen house, although there was a chance that the wives might be out together; or, perhaps, in together. Maybe he could get two at once, although more than one might be too hard to handle. In that case, he would need to make a choice.

30 LUNATIC ON THE LAM

The door bell rang, and Mary Christoffel's first thought was that Rick had once again forgotten his house key. Mary had grown accustomed to the idea that having total recall of the minutest physics concept or some abstract computer syntax did not predispose a mind to remembering something as mundane as a house key.

It was only 4:00 PM, and with Rick's early morning hours at the Message supercomputer, he would often come home at about that time. However, it sometimes meant that he needed a quick fueling for another late night at the computer. In that case, she might be able to talk him into a restaurant meal.

She was surprised when the person at the door wasn't Rick, but rather a scruffy, thirty-something man; and, he was holding a gun.

"Inside," he said, and Mary complied. Mary, always cool under pressure, had her own rejoinder.

"What's this about?"

The man stashed the gun into his pocket and looked somewhat disgusted that he wasn't recognized.

"I guess you don't follow the news. They call me the "Prime Lunatic," but I prefer your calling me Jake."

"OK, Jake. So, this is about The Message?"

"Yeah, yeah, The Message! The unholy message that's gotten me into a lot of trouble. Here it was, I was only trying to help... To preserve the world from that evil! A year ago, you and I didn't even know the other existed, and now our fates are interlocked by that evil!"

Mary was starting to get nervous. There was no longer a gun pointed at her, but this Lunatic was starting to sound as if the name was appropriate. She needed to cool things down.

"Jake, let's sit, and you can tell me why you're here."

"Sorry, no time for that. You and I are going to a safer place. Do you have a coat in the front closet? It'll be chilly where we're going. And get your wallet. We'll need some cash."

He seemed to be considerate, for a lunatic. Mary walked to the hall closet and grabbed her coat.

"My wallet is in my purse. It's upstairs, in the bedroom."

"OK. Lead the way, but walk real slow. I don't want any doors smashed in my face, or anything like that. I'm being nice to you, so be nice to me. Remember, I've got a gun."

Mary walked up the stairs. The bedroom was on the left, at the end of the corridor, and her purse was on a chair by the door.

"Just the wallet," said Jake, and he grabbed Mary's forearm. "Nothing else."

"But my keys. I want to lock the house."

Jake thought for a second, and he figured she would be easier to handle if she were less worried.

"Yeah. Take your keys."

Mary unzipped the purse with one hand and removed the wallet and her keys. Jake kept his grip on her as they walked down the stairs and out the door. Mary stopped, Jake released her arm, and she locked the door. There was a very old SUV parked near the entrance. Jake motioned Mary towards the driver's side.

"Get in. You'll drive."

"I can't drive this! I have a hard enough time driving my own car!"

Jake thought for a second.

"We'll take your car. They're probably looking for my SUV, anyways; but they don't know anything about your car, yet."

Jake had already unloaded the contents of the SUV into the cabin, so that wasn't a problem. Also, the extra gas mileage of the smaller car might be beneficial. In retrospect, using another car should have been Jake's plan all along, but this whole venture was unknown territory to him, so it was a work in progress.

Mary took the driver's seat of her own car as Jake sat next to her.

"Seat belts," said Jake. "We won't be getting stopped for some minor infraction. And, keep to the speed limits."

As Mary pulled out of the driveway, Jake directed her north on one of the interstates. Traffic was slow because of rush hour, but Jake didn't seem too impatient to get to wherever they were going. Mary thought that he needed the time to hatch a plan. Perhaps he didn't know where they were going.

They were about forty minutes out of the city when Jake directed her to an exit to a parallel northern highway. Signs of civilization were becoming sparse. Perhaps an accomplice was waiting in a farmhouse, somewhere. Mary's hopes of being located soon were diminishing by the mile.

Were it not for the present circumstance, Mary would have enjoyed the ride. She hated the incessant traffic in her area. Here, miles from civilization, there were few cars, the trees were a nice change from block after block of building facades, and her mind wandered. Perhaps in the future, when Rick was semi-retired, they could own a nice home in such an area. Coming back to reality, Mary wondered what the near future would hold.

Since there was little need to drive attentively, Mary studied her kidnapper from the corner of her eye. He seemed more nervous than dangerous, and he resembled quite a few men she had seen on campus during her college years. These were "loner" types, never in the company of other men, and certainly not with any women. They were very casually dressed, although not slovenly, and they always seemed to be in a hurry to get somewhere. This was the type who never carried an umbrella against the rain.

Mary wondered how long this drive was going to be. She would soon need a bathroom break.

31 LUNATIC'S LANDING

Rick arrived home a little before 5:00 PM, and he was surprised to see a very old, very rusty SUV parked in the driveway where Mary's car should have been. He wasn't that concerned. One of Mary's many friends was likely having car trouble, and Mary was driving her home after her visit. With a car like that, it was easy to imagine some sort of car trouble. He thought their friends were somewhat more upscale than that!

Rick amused himself on the computer for a while, and he was curious why Mary hadn't yet returned home. He grabbed his cellphone and punched her speed dial number. A few seconds later, he heard Mary's cellphone ringing in the kitchen. Mary was not one to go anywhere without her cellphone!

He grabbed her phone and manipulated the keypad to view her call list. Her last call was very early in the morning. Not surprisingly, it was to Tom's wife, Connie. Rick rang her up looking for some clues. Connie answered.

"Mary, I've been wondering where you were. Have you checked your calendar? Are you free for Friday?"

"Connie, this is Rick, calling on Mary's phone."

"Oh, Rick. Where's Mary?"

"I was hoping you could tell me. Her car isn't here, and there's an old SUV parked in the driveway. Do you know anyone with a very old SUV."

"Sorry, Rick. All the soccer moms we know have late models. There wasn't any note?"

"No, no note. When Tom gets home, have him call me."

They signed-off, and Rick's mind started racing ahead to what he should do. At what time should he consider her to be missing? He needed Tom's advice.

Tom called back right away.

"Rick, Connie tells me that Mary is missing?"

"Well, missing in the sense that I don't know where she is. Her car is gone, she left her cellphone, and there's an old SUV parked in the driveway."

"Did you call the police."

"Yeah, good luck with that! They'll probably tell me that a person isn't really missing until they've been gone for twenty-four, or forty-eight hours. What should I do?"

"I'm close to home, now. I'll get Connie and we'll be right over; but, first, I'm going to call DiGiuseppe and see whether he can do anything."

After Tom's call, Rick searched through the house in the hope that he would find some clue to Mary's disappearance. He found her purse, opened in the bedroom, and her wallet wasn't there. She obviously had her car keys.

Rick's cellphone rang just a few minutes after his call to Tom.

"Rick, this is Joe DiGiuseppe. Tom called to say that your wife seems to be missing and there's an old SUV parked in your driveway."

"Yeah. Mary's car is gone, her cellphone is here, and there was no note."

"Rick, would you get me the license number of the SUV?"

"I'm walking outside now. Here it is."

Rick read the license plate number and heard some typing on the other end of the call.

"Got it. I've give it to you straight. The SUV is registered to the Prime Lunatic."

"What! Why would he want Mary?"

"He's on the run. Maybe he needed a bargaining chip. Perhaps Mary was just an accident. Do you know the license number of Mary's car?"

"I don't even know my own, and Mary's registration would either be in her wallet, which is gone, or in the car."

"No matter. Just give me your address. That's all I need."

Rick gave the address as Tom and Connie arrived.

"Rick, put Tom on the phone, so I can fill him in."

Rick passed the phone to Tom as they entered the house.

"Rasmussen, here."

"Tom, this is Joe, again. The SUV is registered to the Prime Lunatic. Under the circumstances, all of you might be at risk. Lock up the house. There will be some FBI agents there within a half hour."

Tom hung up the phone and turned to Rick.

"Rick, we need to secure the house, in case this lunatic has friends. Joe says that there are some FBI agents on the way. Let's get to work. Connie, you and I will make certain all the windows are locked and we'll close the curtains. Rick, you know best where all the doors are and how to lock them."

They moved in silence around the house, Rick finishing first. They met back in the living room and sat for a few seconds in silence. Rick was the first to speak.

"Why us, and why Mary? It's not like we're at the front lines of The Message. We might as well be technicians, for all we do."

"It's just that we were the closest," said Tom. "It's good that it's nothing personal. And these Lunatics don't seem to be a violent sort. They're like intellectual cultists, online nutbags, you know the sort. This guy is at the end of his rope, and he needed to make some sort of statement. This will get him press coverage, that's for sure. Mary's probably watching television in one room while he's writing some sort of manifesto in another. Where's Mary's cellphone?"

"It's in the kitchen."

"There might be a call on her phone, or maybe on yours. Let's make sure we know how to get her phone into speaker mode, so the FBI guys can hear the call."

Rick fetched the phone from the kitchen, fiddled with the menus, and sat again.

"I guess we just wait," said Connie. "First for the FBI agents, and then for a call."

They didn't need to wait long. The front door bell rang, and Tom went for the door. He peeked through the curtains to see two business-suited men who didn't match his Lunatic stereotype. He cautiously asked through the door who it was.

"FBI," was the short, but effective, reply, and Tom opened the door. The agents flashed their badges and asked whether they could enter, a question that seemed strange under the circumstances, but it was probably a routine part of their training.

Tom led the agents into the living room and introduced Rick and Connie. The agents introduced themselves and got right down to business. Agent Marx, the older of the two, was apparently in the lead, and he addressed Rick.

"Ordinarily, we would move you three to a safe house, but Jake, the Lunatic, might call your landline, so we need to stay here, at least for now. If nothing happens, and it's getting late, Agent Williams, here, will accompany Doctor and Mrs. Rasmussen to their house, where two other agents will be waiting. Likewise, additional agents will arrive here."

Marx pulled out a ragged notepad, which was an antiquated, almost humorous, way to make notes in today's computer age.

"Doctor Christoffel, let's start with a few questions."

The questions were obvious – "When did you first realize..." – etc. Slowly, they worked up to what Rick thought were more useful topics, like the make and model of Mary's car.

"Director DiGiuseppe is handling the plate numbers, so we can start a search for the vehicle, but is there anything about the car that might aid in its identification, like bumper stickers, that sort of stuff?"

"No," was Rick's answer.

"This is an older vehicle. I don't suppose it had one of these roadside assistance communication devices, did it?"

As he was saying no, Rick's face lit up.

"No, but I installed something that might help us! I installed a device in Mary's car to remotely monitor performance. This was in my grad school days. Mary's not that good with cars, and I didn't want her on the road if something wasn't quite right, so there's this thing that plugs into the engine control system. It's a hobbyist device, not commercial. I can query it with my cellphone!"

"Does it have GPS?"

"Well, those things did have GPS, but it was disabled for privacy reasons. If I know how these things work, it was probably just disabled in the software. All the hardware would still be functional. Hardware is cheaper that way, one size fits all. Let me check it out."

Rick retreated to a laptop computer as Marx questioned Tom and Connie. In a minute, he had a telephone number, which he dialed. It wasn't that late in the day, and he was sure this was a one-person outfit, so late nights would be the rule, not the exception. There was an immediate answer.

"Hi! You've reached Autobahnalia. This is Jeff. How can I help you?"

"Hi, Jeff, this is Rick Christoffel. You won't remember me, but I'll be in your database. I bought one of those ECU-cellphone boxes a few years ago."

"Yeah. Those were a big seller. I guess I sold almost a thousand. Have you seen the web site? We've got a newer unit, now. It does a lot more, and it's easier to install."

Rick paused for a second, wondering how he could phrase what he needed to say without too much emotion.

"Jeff, here's the deal. There are some FBI agents here who need to locate the car with my box. I know that the boxes had GPS, but it was disabled. Was it disabled in hardware, or software?"

Now the silence was on the other end of the line.

"FBI? Can I ask what's up?"

Rick got the courage to explain.

"My wife has been kidnapped. She's in the car."

Jeff didn't know how to respond to that, so he just answered the question.

"It was just software."

"As I remember, software patches could be uploaded wirelessly. I did a couple, myself. Can you give us a patch?"

"Yeah. That's easy. But don't fault me for asking this. I need some sort of proof that what you're saying is true."

"Sure, I would expect that. Do you have video conferencing on your computer?"

"Yeah."

"Fire it up, and give me an IP address."

Jeff complied, and within a minute they could see each other.

"OK, Jeff, I'm going to put the FBI agents on the line. Wait a minute."

Rick went into the living room and explained what was happening. Marx and Williams went with Rick to the computer, and Rick introduced them. Marx and Williams flashed their badges for the camera, and Jeff asked them to hold them still for a second so he could get a screen capture for his records.

"OK," said Jeff. "This is either bona fide, or a very elaborate hoax. In any case, no one would fault me for caving in on this. Give me your email address, and I'll send you the patch. It shouldn't take more than a few minutes to dig up the code. I'll also send some instructions on how to upload the patch, in case you've forgotten. Here's my cellphone number if you have any problems. If I don't hear from you in an hour, I'll presume all has gone well. And... Good luck."

Now that things were in motion, the next problem was whether there was cell reception where the car was located. Perhaps Jake, the Lunatic, would call before that happened, but Rick had an interim idea, and he addressed the FBI agents.

"If I give you the device's cell number and ring it up, would you guys be able to get an approximate fix on its position before we do the GPS thing."

Marx answered.

"I think Director DiGiuseppe would be able to handle that better than we can. Williams, get DiGiuseppe on the phone. What's the number?"

"Tell DiGiuseppe that I'll need to call my box. The box will disconnect if it doesn't see my authorization code. Once I'm connected, I can keep it on line for as long as I want. That should make it easier to trace"

⊢⊢⊢⊣⊣⊣

Out on the road, Jake and Mary were making a pit stop at a small, two-pump gas station on a regional road. They had been off the interstate for about fifteen minutes; but, Mary could often spy the interstate in the distance, so the interstate must have been constructed as a replacement for the road they were on. There's only one best way to get between popular populated points.

Mary had been trying to memorize the various road signs and landmarks they had passed, in case that was useful information at a later point. Jake seemed much more nervous than she. As long as their situation didn't change, somehow, she didn't feel threatened. In fact, she was glad that Jake abducted her, and not Connie. Connie would not have handled things as well.

Full service was the routine in these rural areas. In fact, the antiquated gas pumps didn't even have a credit card slot. Payments were handled personally. As they drove to the pumps, Jake reminded Mary that he had a gun, and the gun wouldn't be just her problem, but the problem of whatever teenager or old man was in charge of the pumps.

At the pump, Jake rolled down his passenger window, and asked for a fill-up.

"Cash or credit?" was the reply.

Jake said, "Cash," and then rolled the window back up. The old guy started the filling and then retreated to a folding chair at the entrance to the station building. Unlike a youngster who would be fingering a cellphone, the old guy just sat, watching the cars go by. Jake thought about how uncomplicated this man's life had been, compared to his, even before The Message. Jake had done some interesting things in his life, but what would it be worth after another ten years?

The pump shut off, and the old man walked back to the car. He topped off the tank to the next nearest dollar and went back to Jake's window on the passenger side. When Jake rolled down the window, the old man said, "Thirty-two." Jake had two twenties ready, and the old guy excused himself to go inside for change.

Jake thought about telling him to keep the change, but an eight dollar tip on a thirty-two dollar bill would have been unusual. The fewer red flags on this trip, the better. At this point, they were just one of many tourist cars on the road, so they waited the minute, or so, for the guy to return. When he did, Jake slipped him a dollar, quickly rolled the window back up, and instructed Mary to continue in the same direction. At that point, they had been traveling for about an hour and a half.

After they had been on the road again for a few minutes, Mary got the courage to ask a question.

"How much farther? I don't like driving at night, and I'm getting close to needing a restroom stop."

"Don't worry about that," said Jake. "We're more than half way there, so we'll arrive long before sunset."

Mary, emboldened by the fact that Jake answered that question, tried another.

"Is there anyone there, waiting for us?"

"No more talking. Just drive."

Jake was also glad they would arrive before sunset. His GPS unit, the one that helped him find the side road to the cabin, was still in his SUV. He needed to navigate from memory. He had erased the cabin address in the GPS, so there was no obvious trail there. The cabin address might be lurking somewhere, hidden in the unit's memory, so he wished he had remembered to take the unit. He was too nervous at the time to think straight!

Even if the cabin address was somewhere in the GPS, it would probably take a good hacker a few days to dig it out. Jake was hoping this adventure didn't last more than a few days. The last address was the Christoffel house. Maybe the cabin address was over-written by the Christoffel address. Maybe only the manufacturer had that capability to extract deleted information like that from memory. Jake laughed to himself. The GPS was probably made in China!

⊢⊢⊢⊣⊣⊣

A team of FBI technicians had arrived at the Christoffel house, and they were examining the SUV for evidence.

They had examined the GPS unit and had found only the address of Jake's apartment, which they already knew, the Christoffel address, and a couple of restaurants and movie theaters. They bagged the unit for electronic forensic analysis.

Aside from the GPS, some food wrappers, and an abundance of Jake's fingerprints, that was all they found. It was likely that they found so little because Jake had few possessions. The SUV was towed out of the driveway in less than an hour.

During all that, Rick was working to update the software on the monitor box attached to Mary's car. Connie had made coffee for Marx, Williams, Rick and Tom. It was strange for her to be there without Mary. She glanced at the clock almost every minute, and occasionally at the telephone. Kidnappers always call, don't they? If this guy was a killer, he would have shot Mary right in the house, and then shot Rick when he arrived home. And, it could just as well have been her, rather than Mary!

Rick had established contact with the monitor box in Mary's car, and the update appeared to be successful. The cellphone reception must be spotty wherever Mary's car was at the moment, since the update took far longer than expected. Rick was hoping that there would be reception wherever the lunatic was taking her, or else they were out of luck. He called Marx over to the laptop.

"I've updated the software. Here's where we see whether it works. Luckily, it's all text-based, so the old application on my cellphone will pull the GPS data on Mary's car without any problem - At least, that's the hope."

Rick's hands turned cold as he launched the monitor application. "This *has* to work," he thought as the control panel popped-up. He punched the screen button and hoped for results. There it was! GPS coordinates for Mary's car! He wrote the numbers onto a slip of paper and passed it to Marx, who got right onto his cellphone.

Rick, in a parallel effort, pulled up a map and satellite image of the area. There was a road, but no nearby buildings. It looked as if they were still on the road. Rick called Marx, who was still on his phone, and updated him on the area. These same computer things were likely happening on the other end, perhaps with more sophistication. Marx handed the phone to Rick.

"Doctor Christoffel, how often can we get position updates?"

"I can do it as often as you need."

"Give me another update, now, so we can see whether they're moving."

Rick complied, and when the numbers had changed, it was obvious that Mary's car was moving on the road.

"OK. They're in a rural area, up north. We're not going to call in any local police. It might be dangerous at this point. You understand that it will be better if we let them get to wherever they're going?"

"Yes, you're right. I'll monitor their position, and we'll give you updates along the way."

Rick passed the phone back to Marx, who talked for a few minutes more. When he closed the phone, Rick asked, "What now?"

"Well, this might be easy," said Marx. "We're sending some agents into the general area you pin-pointed, just to be on hand. This guy obviously won't make a call from wherever he plans to go. He knows how easily he could be traced. He would travel a distance, maybe even a half-hour drive. Since it appears that his only transportation is your wife's car, he'll take that, and we'll know he's on the move."

Marx continued.

"When we see that he's safely away, we'll rescue your wife. Then, when he returns, we'll get him. Nice and easy."

Rick could see the logic in all that, except for one thing.

"What if he brings my wife with him for the call?"

"Yeah, that could happen, too. If the place is empty, we'll change our plan."

Marx didn't elaborate on what the change might be.

As the evening set it, so did hunger pangs. Tom telephoned for some pizza, and he cautioned the FBI men not to shoot the delivery boy when he arrived. When 8:00 came, two other FBI men arrived at the house to relieve Marx and Williams. Marx and Williams followed Connie and Tom back to their own house where other agents were patiently waiting in a car, parked on the street.

The first thing that Tom did when he was home was to arrange cover staff for both Rick and himself at the supercomputer. Rick, of course, was out for the duration, and Tom intended to remain with him through this adventure.

Tom and Connie had nothing better to do, so they headed to the bedroom shortly after ten. Tom had instructed his FBI house mates on the location of the bathroom, how to use the coffeemaker, and the proper use of the television remote. The agents didn't seem too concerned that anything would happen, but Tom was also certain that they wouldn't be doing any napping on the job.

Going to bed was one thing, and going to sleep was another thing altogether. He suspected that Rick was still pulling position updates from his cellphone. He would likely continue until it was certain that Mary's car had reached the Lunatic's intended destination. Then what? He had heard Marx's game plan. It sounded all too simple, but he wouldn't tell that to Rick.

He had watched some of those crime shows on television. SWAT teams, hostage negotiators - How was this going to play-out? What would this Lunatic say when he finally telephoned? What did he want? Tom didn't think this was about any ransom money. Safe passage to some South American country without a Lunatic extradition treaty? He laughed to himself when he thought of a possible newspaper headline - Lunatic Asylum.

32 THE CABIN

As they neared the cabin, Jake tried to formulate a plan. He had to admit that at this point he didn't really have a plan, he was just reacting to events. Nevertheless, he did have an objective. He wanted the world to realize that The Message was evil, that its reception should be stopped, and any decoding efforts should be stopped also. He somehow needed to formulate a plan around the objective.

Before he abandoned his apartment, he took the precaution of buying four prepaid cellphones. To allay suspicion, he bought two each at two different stores. He had paid cash, as if that really mattered. He was a marked man in any case. When he made calls, they would be a single call from each phone, and each phone would be destroyed and quickly discarded after use. He was smart enough to stay one step ahead of the cops.

Their road trip to the wilderness had been happily uneventful. After their fuel stop, they did pass a police car on the road. A description of the automobile and their license plate number apparently hadn't been posted to any "all points bulletin" web site, since the cop didn't react in any way. Jake laughed to himself that the cop must have been out on a donut run, and his only thoughts were about the greasy goodness that awaited him.

As Mary reached the approximate area of the cabin road, Jake had her slow down, so he could look for landmarks. He was happy that there was no other traffic on the road, since he didn't want any stray clues about their location.

"Go real slow, now," said Jake, when it looked like they were approaching the road. "There, take a right there." He pointed to the barely defined dirt road leading to the cabin.

Mary stopped the car. "There? Where?"

Jake tried not to answer too angrily. He needed a peaceful captive. As it was, he didn't like the idea that he would need to tie her up so he could leave to make calls.

"Look for the tire tracks, just twenty feet ahead."

Mary had the car crawl forward, she saw the tracks, and they turned in. The road, if that's what it could really be called, was very rough, so she proceeded dead slow. After about three minutes, they reached a small cabin, very isolated, and very hidden in the woods. At that point, Mary was praying for some indoor plumbing. It did look like the place had electricity, so maybe there was a water pump.

Mary stopped the car in front of the cabin and shut off the engine. She removed the keys and was in the process of pocketing them when Jake said, "I'll take those." Mary handed them over, and they exited the car. Jake motioned for her to grab her coat and precede him. When they reached the door, she stopped, waiting for Jake to unlock it. The door was unlocked, and Jake just opened it and motioned her through.

The cabin was actually quite nice. There appeared to be a large living area combined with a kitchen, and two doors. One door likely was the bathroom, and the other, a bedroom.

"I need to use the bathroom," said Mary.

"That door," said Jake. "There's a window, but don't even think about using it. You'll never outrun me in these woods, and I've got the gun."

This certainly wasn't a cabin occupied at any time by a woman. The bathroom hadn't been cleaned in quite a while. At least there was a full roll of toilet paper and soap at the sink. There was just one faucet, cold, and the water was very cold. This must be a hunters' or fishermen's cabin, with a few modern conveniences, such as electricity and indoor plumbing. The men, however, still wanted to think that they were "roughing it."

When Mary left the bathroom, Jake was sitting on a chair that seemed to be made from tree logs. He was writing something. Mary didn't know what to do, so she sat on a dilapidated couch in the center of the room. As exciting as this might be to Jake, Mary was utterly bored.

Jake stuffed the paper into his shirt pocket and turned to Mary.

"I'll need to leave for a while. Put on your coat."

Mary complied.

"I'll try to do this as gently as possible, but I'll need to tie you to a chair."

Jake pointed towards the chair he was just sitting in. It was an obvious choice, since it was too heavy for Mary to move.

Mary sat, and Jake unwrapped a package of some freshly purchased rope. He held the rope for a few seconds, not doing anything. This was something that he definitely didn't want to do. Finally, he knotted the rope onto one of the rear chair legs and swung it around to the front. He pulled the sleeves of Mary's coat beyond her hands and tied the sleeves together. That solved the problem of her hands.

Jake worked methodically around the chair, looping the rope through the openings in the chair and around her body. He pulled the rope tight, but not as tight as he could. When he was done, he tested the rope, here and there, and addressed Mary.

"OK. That's done."

Jake looked relieved. This might have been as close as he had ever been to a woman. If Jake had any bondage fantasies, they certainly didn't show. Although Mary was the victim, she felt sorry for Jake.

"I'll be gone for about an hour. When I get back, we'll eat. I'm trusting you not to struggle. I'll know if you tried when I return. Just sit still, and it'll be easier on both of us."

Jake left, not locking the front door. Mary got the idea that he didn't have a key. He might not think that locking the place was necessary. They were in the middle of nowhere.

⊢⊢⊢⊣⊣⊣

At the Christoffel house, Rick was pulling the GPS coordinates from Mary's car at about five minute intervals. Since the car had been stationary for quite a while, Rick pulled up a satellite image of the location. He could see nothing but trees, and the place was quite a bit from the road. He couldn't imagine how Mary's car could have driven there.

Mendez, one of the second-shift FBI agents, passed the GPS coordinates over to someone on the other end of his cellphone.

"What do you think?" asked Rick.

"This might be it," replied Mendez. There's no telling what's in those woods. Maybe just a tent. Unfortunately, this is a remote location. Our team is still about a half hour away. I'll tell them to step on it."

Rick went for a wider map image. The area was rather desolate. It looked as if there was a gas station or convenience store about five miles from where Mary's car was parked. The Lunatic wouldn't chance a call that close. Going very wide, Rick searched the highway in both directions out to a fifty mile radius. He pinpointed a few locations where there might be pay telephones, although that would be a long shot. In the age of cell phones, who needed a pay phone? The Lunatic could call from anywhere he could get cellphone reception.

It was just fifteen minutes after Rick noticed that Mary's car was parked that it went into motion again. It was back on the highway about five minutes after moving. Mendez was having a running conversation with someone on his cellphone. He overheard Mendez ask, "Where are you now?"

Mendez asked Rick to pinpoint the car on the map, and he continued his cellphone conversation.

"He's going in the opposite direction from where you're coming, so no problem there. We'll keep you updated."

Rick looked up at Mendez, hoping for an update.

"So far, so good. Our team is approaching from the other direction, so they won't pass the car on the road. The vans are unmarked, but there isn't that much traffic out there. We don't want any suspicions, especially if Mrs. Christoffel is in the car. Two identical vans traveling together would be an anomaly in that area."

"How far away are they?" asked Rick.

"About fifteen minutes. I think we have a good safety margin. Keep monitoring."

While staring at the computer monitor, Rick's mind turned to how things might have been. Mary wouldn't have been in such danger if he and Tom had not become a part of the Message team. In the road not taken, they would now be seeing the first successes of their business venture, still dreaming about a secure future and an early retirement. His work location had not changed, he still lived in the same house, but everything was now so different.

To Tom, it was just another way for the company to make money, this time with fewer financial risks. Rick's enthusiasm for scientific adventure made it easy for him to become a part of the Message team. He might only be a footnote in the history books, but he would *still* be in the history books. Their financial venture was just one of many of their age, and it would be forgotten among the rest.

Connie and Mary had welcomed the financial security of Rick and Tom's company becoming a long-term government contractor. It was time for the group to settle down into a routine. It was ironic that Mary's support of the change in business plan was motivated by security, and she ended up being kidnapped.

When this whole episode was over, the Message people would surely give them better security. It was the same locking the barn door after the horse escapes scenario you always find with government. The government only reacts to situations. If it were more analytical, it could save many lives and a lot of money.

Mary's automobile was still moving, which was good news for the FBI agents, since they were assured of reaching the place where Mary might be long before her kidnapper returned.

There was still the problem that there might be an accomplice with Mary, or even a band of kidnappers. Although the agents were fairly certain that the Prime Lunatic was acting alone, they assured Rick that they would take no chances in Mary's rescue. If she wasn't at the wilderness location, then it was presumed she would still be in the car. They would back off and wait.

Finally, the position of Mary's automobile held constant for a few minutes at quite a distance from where Mary was presumed to be held.

33 RESCUE

Jake set out to make his first call. Back at the cabin, he had carefully crafted his demands; or, rather, the demands of the Lunatic Fringe. His fellow members were not consulted in the content, but he was certain there would have been no objections to any of the demands. These demands had been their goal since The Message had been announced. The words flowed easily from his mind to paper, but he checked and rechecked the wording for clarity. Jake was a man of few words, and he didn't want any of those words to be misunderstood.

Jake carried a disposable cellphone in each front coat pocket. He needed only one, but he was trained as an engineer, and he always looked for second-order effects. If one cellphone didn't work, he didn't want to drive back to the cabin for a replacement. It had been too many hours already since he abducted the woman, and everyone was expecting a call.

Jake didn't think his cheap throwaway cellphones had any GPS capability. Any call, however, could be crudely traced by triangulating its position to multiple cell towers, so he decided to drive quite a distance from the cabin to make the call from some remote parking lot or rest stop. He was happy that this rural region had very few cell towers, so triangulation would be difficult. Triangulation would be impossible if his call was short, so he planned to read his statement, say he would call back in twenty-four hours for a response, and then smash the cellphone. He was smarter than the average kidnapper.

Under different conditions, Jake would have enjoyed this nighttime drive. The sky was clear, with a slender crescent of a moon and many visible stars. Jake wondered from which of those stars The Message was beamed. Some of the Fringe called it the "Demon Star," but Jake didn't see things that way. The star and the senders of The Message weren't at fault. It was humanity and how it would twist The Message to its usual evil purposes that's the problem.

He was happy that the tiny car was getting good gas mileage. The gas gauge had just barely moved from the full mark since they left the old man at the service station quite a while ago. Jake had been traveling for almost half an hour, so it was time to look for a convenient calling spot. He passed a 24-hour diner whose parking lot could serve in a pinch, but he was hoping for someplace deserted.

That's when he came upon a transmission fixit shop, naturally closed at that hour, but with a parking lot full of cars in which his car would blend nicely. He pulled in, turned off the lights, waited to make sure he really was alone, and grabbed a cellphone. Reception was marginal, but workable. He carefully reviewed the text of his demands, swept his shirt pocket for the slip of paper with the woman's home telephone number, punched it in, took some calming breaths, and hit send.

⊢⊢⊢⊣⊣⊣

The FBI vans arrived near the cabin shortly after it had been verified that Jake was on the move. Their immediate problem was that Jake might have an armed accomplice still in the cabin, so they used their GPS to stop a few hundred yards away and out of sight. A team of six men in full SWAT gear quietly approached the cabin. Seeing that there was no rear door, they merged at the front.

After settling into position, three of them rushed the cabin. They kicked down the door and prepared to disarm any occupants. Instead, they found Mary, alone, tied to the chair where Jake had left her.

Mary, startled, could only say, "That was fast," as one of the agents untied her. He handed her a cellphone so she could call home. Not surprisingly, her call was answered immediately.

"Rick, it's me. I don't know how anyone could have found me, but I've been rescued."

Rick could hardly talk, but he managed to get a few words out.

"It was your husband to the rescue, with an assist from the guys you see there, but I'll explain later. And the guy..."

"Well, the Lunatic left to make his call, and that's when I was rescued. He's still out there, and..." Mary paused, not knowing how to say it, "I'm worried about him. He's not that bad of a guy, kidnapping me notwithstanding, and I wouldn't want him to get killed."

"Does he have a gun?"

"Yes, but he's uncomfortable with it. He never really harmed me, but these guys look like they'll shoot him on sight."

"Mary, let them handle it. They'll give him his chance to surrender. If he's smart, he'll be OK. I just want you back home safe with all this stuff behind us."

Mary handed the phone back to the agent and sat back down. Another agent, who was apparently in charge of the operation, walked over to her.

"Mrs. Christoffel, I need to ask you some questions. We can't get you out of here just yet, since your abductor might arrive at any minute. Put this on."

He handed her a bullet-proof vest and helped her fasten it.

"First, was he alone?"

"Yes."

"Does he have a gun?"

"Yes, but about that..."

"No time, that's all I need to know. Sit well to the back of the room... That couch looks fine... And we'll guard the door. There are two others outside, and another who's moving our van out of the way, so he won't know we're here."

⊢⊢⊢⊣⊣⊣

From his own viewpoint, Jake's call had gone well. He had stuck to his script and didn't allow anyone's keeping him on the cellphone long enough for any sort of trace. He smashed the phone immediately, then he threw the pieces into the woods where they wouldn't be discovered.

The entire venture was playing back in his head, over and over, as he headed back to the cabin. His mind wasn't on the road.

"Damn! Missed the turn! Now, I'll need to find a place up ahead to turn around."

As Jake drove past the cabin turn-off, well hidden in the night, he passed two vans, parked alongside the road a few hundred feet ahead. Their lights were off; so, if they were occupied, whoever might be in them wouldn't quite recognize his vehicle or read his license plate. Jake, however, could see that the plates on one van, so incongruously parked in the middle of nowhere, did not look like standard issue plates. His mind raced as he kept driving.

"Could they have found me so easily? If they didn't, then I can't leave that poor woman tied up there." Jake didn't know what to do, but he did know that he shouldn't go back to the cabin. He sped away, hoping to get some ideas while driving. At first keeping the speed limit, so not to attract attention from the vans' possible occupants, he then sped off into the night.

⊢⊢⊢⊣⊣⊣

Back at the Christoffel house, Rick was still pulling GPS coordinates for Mary's car. When it was close to the cabin, one of the agents telephoned a heads-up to the cabin team. Then, the car passed the cabin and kept going. Rick was confused.

One of the agents at the house saw what was happening and had his analysis.

"He must know we're there."

The cabin team was now at a disadvantage. Their suspect was racing down the highway, with a healthy head start, and their vans were several minutes away by foot. They did have the advantage of Rick's tracking, but the

Lunatic might now be suspicious about that and abandon the vehicle.

After a quick consultation, the agents at the cabin devised their plan. One agent would stay at the cabin with Mary while another team arrived to bring her back home. The rest would follow Jake in their vans, but at a distance, and not try to chase him down. As long as they could track him, that was good enough. If he stopped, then they would try to corner him. If he moved closer to civilization, they could enlist the local police in a road block.

Mary was alone in the cabin with one of the agents, and she tried to explain to him that Jake wasn't that dangerous. They should make every attempt to capture him alive. The agent explained that whenever a gun was involved, they had no options than to use their own weapons if threatened. And, after all, didn't that guy just kidnap you at gunpoint?

⊢⊢⊢⊣⊣⊣

Jake initially raced down the road; but, when he saw that he wasn't being followed, he pulled back to the speed limit. Although they may have the woman, they might still have no idea where he was. And, how did they find the woman? He had no idea, but he did know what he would do next.

He still had a cellphone in his pocket, and this time he knew whom to call. Pulling over to the side, he fished Amy's card from a pocket. She had given it to him at his interview, and it had her cellphone number on it. It was early morning in New York, but he dialed anyways. Amy answered on the third ring.

"This had better be good! It's 4:15AM!"

"Sorry, Amy, this is Jake. I have a big story for you."

It took a second for Amy to realize who Jake was.

"Jake, the Lunatic?"

"Yeah. No one else in the press should know, but I kidnapped the wife of one of The Message people."

"What!"

"Happy ending, at least for her. I think she's been found and she's none the worse for the experience. I had hoped to use her as a bargaining chip to stop The Message, but it didn't work out that way."

"You're in jail?"

"No, still free, and deciding that maybe the best outcome of all this is publicity for my cause. That's why I want you to broker my surrender."

"How can I do that? I'm thousands of miles away."

"I'll give you a phone number. Call that, and they'll be interested in anything you say. Hold on."

Jake retrieved Rick's number from his pocket, he read it to Amy, and he had Amy read it back to him.

"Yeah. You've got it. I'll call your cellphone tomorrow at eight AM. That's eight AM, my time, got it?"

"OK. Eight AM your time. What time is it now?"

Jake checked the automobile clock.

"About a quarter after one."

"OK. That checks, you're three hours behind me."

"I just need a safe haven for the surrender, a city park with a lot of people, something like that. I don't think they would gun me down with so many people present, but with those guys, you never know."

"Jake, if you play it cool, hands up, and all that, you'll be safe. I distrust the government types myself, but I don't think they'll do anything like that."

Jake laughed. "Yeah, imagine all the extra paperwork."

"I'll wait for your call."

Jake shut down the phone, took out the battery, then paused for a moment. If he destroyed this phone, he would need another. Checking his wallet, after gas money, there wouldn't be enough left. Damn! He needed to keep this phone! He thought of the two others he had left at the cabin. If only he had brought them all with him. Too late for regrets, now. Without the battery, the phone was safe. Since it was Amy's number that he called, there was probably no trace. He just needed to be careful next time he used it.

Jake pulled back onto the road with nowhere to go. He needed to sleep. Maybe he would find a vacant campsite somewhere around, so he could spend the night in the car. He was a little hungry; but, at this point, food had a much lower priority than quite a few other things. He knew he could trust Amy. At the very least, she was in it for the story.

As he drove along the highway in the early morning hours, Jake had time to think. He thought he was smarter than the average kidnapper; so, how did they find him so soon! There were rumors on the Internet about spy satellites that could read license plate numbers from Earth orbit. There were also those damned drones that law enforcement agents were buying by the dozens. However they did it, it didn't matter. His initial move, the Christoffel kidnapping, was over. He was deep within the middle game, and he needed to plan his end game for maximum effect.

Jake thought back to some of the *film noir* movies he had seen. They were entertaining because the private eye was always several steps ahead of the inept police. Jake's kidnapping would have been much easier in those days. His problem was not that the police had suddenly become smarter, but that they used the technology that he and his colleagues had developed over the last several decades to do their job for them.

In the past, people were critical of "armchair generals" who sat in their luxury offices, never seeing combat. In the modern world, police were sitting in front of computer monitors, looking for their next easy collar. Many of them were just public paid minions of corporations looking for some hapless downloader of a copyrighted movie. The emphasis of law enforcement had shifted to white collar crimes, detected with little effort by digital forensics. White collar arrests were easy, and the headlines that they generated justified an increased funding for more technology.

Jake began to regret his career choice, since it had helped to enable the flawed society in which he lived. At this point, it didn't matter. His career was over. At best, he would be spending most of the rest of his life in prison. At worst, he would be dead within the next twenty four hours.

34 SURRENDER

Amy rang Rick's number immediately. She didn't know whose number it was, or what to expect.

"Hello?"

"My name is Amy Pomronger, a journalist, presently in New York City. I was given this number by Jake, the Prime Lunatic. He wants to surrender."

One of the FBI agents had answered Rick's phone.

"This is FBI agent Mendez. Hold the line."

"Well, forget about the middle-man," thought Amy, "I got a direct line to The Man."

Mendez conferred with his partner, who called headquarters.

"Amy, is it? We're going to give you the number of our section chief. Call him immediately and give him the details."

Mendez gave Amy a telephone number, and before calling, she attached her cellphone to a charger. She reasoned that this would be a long night of calls.

"FBI. Bellisario, here."

"This is Amy Pomronger from New York. You're expecting my call."

"Amy - Can I call you Amy? – How did the Lunatic contact you?

"By telephone."

"Can I have the number?"

"Wait a sec..."

Amy realized that Jake might be tracked via the number, so she decided not to give it out. Either Jake's paranoia was rubbing off on her, or distrust of police was becoming the rule in today's society.

"...Sorry, I can't seem to pull up his call information on my phone," was Amy's response.

"Why did he call you instead of anyone else, and what did he say?"

"Well, I interviewed him a while ago for an article, and I gave him my business card. He probably thought I could be trusted, so he called me. He admitted to kidnapping someone, he didn't specify whom. He said she wasn't harmed in any way and that she was likely in the company of your agents. He figured he made his play, lost, and now he's looking for a peaceful solution. One thing I know from experience, Jake's a paranoid character. It isn't unusual that he fears for his life, even while surrendering. He wants to do it in public."

"Amy, that guy's a criminal, probably a crazy, with a gun. We need to do whatever we have to do."

"Look, if you're going to gun him down in cold blood, then I'll end this telephone call right now! Also, I'm a journalist, and if you do, I'll be sure to write all about this!"

"No, no. If he shows no threat, he'll be OK. Where does he want to surrender?"

"He hasn't decided yet." Amy decided to leave out details about his planned 11:00 AM call. "He'll tell me when he's ready, and then I'll tell you."

Bellisario asked, "Can we suggest a place?"

"Ha! Wherever that is, that's the last place he would want! He doesn't trust you, and I must say from the tone of this conversation, he might be right. One thing I need to know before speaking with him again, whom did he kidnap, and do you really have her back safely?"

"The woman is Mary Christoffel, the wife of one of The Message computer people. She's now safely on her way home, where she'll be under guard until this is over."

"When there's more information, I'll call your number."

Bellisario answered, "Amy, I'm going to send some agents to your place. Where are you?"

Amy wasn't going to hand Jake over to these guys so easily.

"That won't do any good, since I'm leaving as we speak."

"Where are you going?"

"Sorry, my itinerary is fluid, and even I don't know where I'll be at any one time. You'll just need to wait for my call."

Amy shut her phone, and her next move was to make certain that her editor, Andrew, didn't get any sleep.

"What the hell?"

"Andrew, this is Amy. Were you sleeping?"

"Of course I was sleeping! What's up?"

"An exclusive story, hot for the wire services, that will make you and me famous - Well, famous for about a week."

Amy outlined Jake's kidnapping, and they decided she would write a short release, just announcing the kidnapping, but not mentioning Amy's role as the go-between in Jake's surrender. The longer story would appear in their own magazine.

"Andrew, I'm leaving my apartment, now, and I won't be found until after the Lunatic's call. There might be some government heavies poking around the office this morning looking for me. You don't know where I am, since I'm not telling you. Expect my news flash as soon as you arrive at the office."

Amy grabbed her laptop and cellphone, and she headed for the subway. She wasn't driving anywhere for fear of being tracked in her car. Her destination was a coffee house, followed an hour later by another, etc., until Jake's call. She removed the battery from her cellphone, so she couldn't be tracked. She would replace it shortly before eleven. Amy intended to file her story remotely from the coffee shop. It would take just a few minutes to write.

Andrew was too pumped from Amy's call to get back to sleep. He showered, grabbed a peanut butter sandwich for breakfast - he had been eating like that since college - and he headed to his office to await Amy's story. The magazine rarely got items newsworthy enough for a wire service release, so he re-read the guidelines on how that should be done. Fortunately, it was an editor's decision what to release, so he didn't need to worry about any approval. Even management was willing to concede that the news travels too fast for that kind of oversight.

⊢⊢⊢⊣⊣⊣

Jake made certain he kept to the back roads, and when he was an hour's journey from the cabin, he started to look for a place to bed down. Fortunately, he was still in the wilderness, and he found a dirt road leading off from the highway. He didn't know where it led, but he pulled over to the side, a few hundred feet in, where he wasn't visible from the highway. Unless this was some rush-hour short-cut in these parts, he would be undisturbed until morning.

He still needed to decide where he would surrender. Someplace like a shopping mall might work. He knew a few in some suburban areas that would do the trick. There was one for which he could park at an adjacent big box store instead of the mall lot. In that way he would avoid being apprehended in the parking lot, though they would likely have agents stationed at the doors. No matter. He would mix himself in with a crowd of shoppers, and they would act as his shields so he could surrender in safety.

Unknown to Jake, who was still in Mary's car, he had been tracked the entire way, and the FBI hadn't signed onto the same plan. When the car had stopped, the agents in the following vans were informed, but they had one problem. The automobile tracking coordinates were precise to about a hundred feet, so when they got really close, they needed to tread softly.

They passed the dirt road, and parked a short way up the highway. There was some moonlight, but not enough to slog through the forested area in search of Jake's car. There was also the possibility that Jake wasn't in the car, and that there might be another safe house, perhaps with some accomplices.

They took Jake's overture for a surrender as genuine, re-parked one van across the dirt road, disembarked, and waited in the woods for morning and his likely exit. If he did mean to surrender, whatever happened would happen there, not in a public place.

Amy had a hard time staying awake, even after the multiple doses of caffeine from her several coffee house stops. Maybe she could parlay this experience into a review of New York coffee shops. She had emailed her story to Andrew using her cellphone, not the coffee shop Wi-Fi, and it was nearing eleven, which was Jake's specified calling time. Amy once again plopped the battery into her phone and waited.

Across the continent, Jake was stirring from a half sleep, feeling a need to relieve himself. He hadn't slept that well, since he was constantly checking the time so as to not miss his call to Amy. First things first. He exited the car, then unzipped to water a nearby bush. He had decided on where he would surrender, the mall near the big box store, so he started the engine and simultaneously called Amy.

Amy answered immediately; and, just as immediately, Jake saw the van blocking the road. Jake stopped the car.

"Amy, I don't know how, but I think they're here! What should I do?"

This was not the call that Amy was expecting.

"Jake, put the phone in your shirt pocket, so I can hear what happens. Next, make sure you leave any gun in the car. Last, exit slowly with arms raised and walk towards them."

"Amy, I'm doing what you said. If I'm a martyr for the cause, I want you to tell the world what happened. Maybe then they'll stop The Message!"

Amy could hear the car door open, then close. The agents, who had been half asleep, didn't notice the car. However, they did hear the door. They appeared at the roadway and saw Jake walking towards them with arms raised. Amy could hear a few voices in the distance, then a few louder barked orders, the rustle of Jake's apparently lying on the ground, and then a lot of other noises. Somewhere in there, Amy thought she heard his rights being recited. Happily, no shots were fired, so the surrender went as well as could be expected.

One of the agents noticed the open cellphone in Jake's pocket and asked who was on the other end. Amy told him only that Bellisario would explain, and she exited the call. Now, to her office, to Andrew, and the second news flash of the day. On her way to the subway, as she dialed Andrew with the details, she was already composing the article she would write about Jake's capture.

Jake, handcuffed in a back seat of a van, had time to reflect on the last twenty-four hours. Things had ended as well as could be expected. He was alive, and he didn't need to live through what would have been a stressful few hours of his own surrender plan.

Jake's object in the kidnapping was to get his story out to the public. The kidnapping and his capture would be front page news that night, and he was hoping that his contact with Amy would keep the story going a little longer. From their interview, Amy had detailed knowledge of his cause and his motives, so any story she wrote would give people a better understanding of the Lunatics and the evils of The Message.

Jake had a further hope that his trial would keep the Lunatic cause in the public eye, but he was too savvy to think that any of his motivations for the kidnapping would be highlighted in the trial, itself. The government, and the so-called "justice" system, would proclaim his motivations as irrelevant. For them, it was "Just the facts, Ma'am." There was a kidnapping, he did it, and that was that.

That's why Jake was counting on a fair handling of the affair by Amy. He realized that his interview with Amy would be important for the Lunatic cause. Now, he saw what a fortunate interview that had been. He had Amy pegged as a closet feminist, and feminists had always been good fighters.

The FBI had informally debriefed Mary in the other van on the way home. Under the circumstances, they would wait a while for her formal statement. The FBI agents were surprised that Mary was thinking more about Jake's fate than the personal events of her kidnapping. She appeared relieved when they got word that the surrender had been successful.

35 END OF TRANSMISSION

Mary's kidnapping put everyone on notice that The Message principals needed better security. Although Jake, the Prime Lunatic, was in custody and his organization was presumably in shambles, there might be worse threats out there. Jake's actions were reported as being politically motivated, but some argued that his reasons were more philosophical than political, and there were religious extremists out there who might not be as polite a criminal as Jake. Fortunately, foreign government agents were not considered to be a problem, since the Message Consortium was international.

Against the advice of the Message Consortium, and also against her husband's wishes, Mary accepted nearly every interview request she received. In all these interviews, she emphasized the fact that Jake had treated her well, and that he had surrendered in the end. She didn't want Jake to be too heavily punished. Amy interviewed Mary, also, as part of the story about her experience as Jake's go-between with the FBI. Although the FBI didn't publish the fact, Mary disclosed to Amy that it was the tracking device in her car that led to Jake's easy capture.

Not surprisingly, tracking devices were installed on all the vehicles used by The Message principals and their families. In all cases, body guards were offered for around the clock protection, but Rick and Tom opted for more secure entry ways to their homes and extensive home security systems. Their security included twenty four hour video surveillance of the exteriors of their homes by a Message contractor. They had panic buttons installed at various places in their houses. One push would cause a flurry of armed agents to descend on their houses.

The message decryption was proceeding as expected, with the low data rate message providing more clues to the identity of The Senders. Other images were decoded to mean that The Senders, too, were on a world with several continents surrounded by large bodies of water; and, that their life evolved from the seas, just as life on Earth evolved. It seemed as if the low data rate message was building to the climax of revealing the body plan of The Senders, and then it happened. There was an abrupt end of the transmission.

The monitoring continued on the chance that this was a temporary interruption. Analysis of the signal indicated that the signal loss was not the result of some object's blocking the signal. The Senders had actually stopped transmission. Roos' first thought was to examine the end of The Message to see whether there was some unusual pattern that might, in effect, be saying "end-of-message," something like a long, alternating string of ones and zeros. The data stream at the end was statistically identical to what preceded it.

There was a joke that The Senders had lost their research funding, and there was also speculation that The Message might have been the last call of a dying civilization that wanted to impart its knowledge before it was gone. Roos was hoping that The Senders were just rotating through reception candidates, and The Message would be back again in a few years.

They did get a complete copy of the fast data rate portion of The Message; but, if the period of the low data rate message was the same as the fast message, they missed more than half of it, and that was the message that had given them the most information. Roos was hoping that similar data to that of the low data rate message would appear in the main message, but it might be harder to separate it from everything else. Still, they were trying; and, even with termination of The Message transmission, there was work for the next few years.

Roos, himself, thought that his formal organization should disband within the next few years, leaving in place the loose network of academics working on the decryption. He reasoned, also, that it would be disheartening for people to work in an organization with the fewer anticipated breakthroughs that would appear years down the line.

One thing that was certain, at least to Roos, was that there should be at least a sporadic monitoring in the event that the signal returned. The technical requirements for just such detection would be simple, so some of the lesser observatories could handle the task. If the signal were again detected, there would be a procedure in place to rapidly acquire The Message as before.

The end of The Message allowed Roos to evaluate his position and his future. If the message-decoding was delegated to a distributed cadre of academics, his role at the Message Consortium would devolve into that of a typical administrator. He might play with The Message himself, but his other responsibilities would get in the way. He would probably be more comfortable on the academic side of the funding stream, and he was certain that he could get a chair at nearly any university he might choose.

As Roos saw it, it was another year at most at the Message Consortium, followed by either his chicken farm, or back into academia. At least he wouldn't be expected to do any more leading-edge mathematics, since his new specialty would be theories for decoding The Message. This might be considered to be applied math, but it was more fundamental than many other mathematical applications.

36 LIFE GOES ON

As Roos had predicted, once the limit of The Message was known, the member nations started an orderly reduction in project scope and funding. They did heed the advice of monitoring The Message frequency to see whether it would once again appear, but the conventional radio astronomers were happy to once again have full control of their telescopes.

The Message Consortium was reduced in size, and it served to provide funding to other organizations for message decoding. Tukey retreated back to his old office and observatory, where he again searched for SETI signals. Only now, there was government funding for the search.

Also, as Roos predicted, he was inundated with offers of university positions. Weighing them carefully in his usual manner, he easily chose the one that was right for him. He sold his chicken farm, but only after returning to it to give farewell bonuses to his faithful employees and to announce to them that a stipulation of the sale was that they would be retained for at least the first two years with their current pay and benefits.

Partly through Mary's efforts, Jake came away with a short prison sentence, but his life was essentially ruined. Irrespective of his being a convicted felon, his engineering profession would be decimated by the time he had served his sentence. Even an engineer more qualified than Jake would have difficulty finding work.

Mary Christoffel had been approached by a publisher to write a book about her kidnapping ordeal, something sure to be a best seller. Instead, she proposed that she and Connie write a children's book about The Message, and her offer was quickly accepted.

It was Amy who wrote a book about the kidnapping, even interviewing Jake in prison. Jake had accepted his sentence with dignity, and he was grateful for the kindness that Mary had shown him during his trial. He had even joked that prison life seemed better than his old engineering job.

Rick's computer was still essential to Message research, so he was persuaded to stay with the Organization. One important task was preservation of The Message for future generations. The digital Message, itself, was just a little more than 28 gigabits, or 3.6 gigabytes, something that easily fits on a single DVD and would hardly tax the storage space of most computer disk drives. There were a myriad number of these copies throughout the world, but Rick wanted something that would assuredly exist for a millennium. That's where his physics training became important.

Rick joined the university team working on the archival data disks of The Message and ways they might be preserved through the centuries. Although copies would be placed at major museums, Rick thought that monuments should be designed and built to contain additional disks.

Archiving the raw data, as received at the radio telescopes, was a harder problem that Tukey and his team was tackling. At this point, the only option was to gather the many reels of tape into a controlled atmosphere environment and arrange a schedule when they would be re-recorded onto newer tapes to prevent data loss. Everyone knew that this arrangement wasn't optimal, and they would just need to wait for a better storage technology.

Tom was still an entrepreneur at heart. He was remaining for the moment, but still looking at business opportunities. Constance made Tom promise to only look for local opportunities, since she was comfortable in her home, and she enjoyed Mary's company.

37 THE PAST AS PROLOGUE

Larchmont answered the communicator at the second twang. The display announced a call from Candle.

"Hey Candle, how are they hanging?"

"Three at a time, like always. So, how's the family?"

"Well, the little guy is causing some problems. Now he's interested in physics, of all things! Maybe it's just some rebellious phase. No one wants those guys around anymore, except for comic relief. Their experiments got too darn expensive. I realize that he's only eighty, so maybe he'll grow out of it."

Larchmont paused for a second, trying to phrase the sentence in a way not to offend. He was too busy that day for idle chatter. Maybe they could schedule a dinner with the wives and talk then.

"Not that I don't enjoy chatting, but is there a reason for your call?"

"Ha! Always the sociable type! I just wanted to congratulate you on that nice job of asteroid deflection!"

"As if we can't do something that simple after all these years. And, don't forget, I had a vested interest in not having that baby obliterate my house."

"Sure, but I was thinking about the way you did it. Brilliant!"

"I can't take all the credit. The wife's an artist. It was mostly her idea. I just supplied the technology."

"Don't be so modest. We've been using radiation pressure to steer those things away from the planet for a couple of centuries, so the whole thing was getting a little boring. Your idea of using radio frequencies to do it was unique. No wonder you got the funding."

"Well, the cost is about the same any way you do it, but I'm happy that the wife's proposal was selected."

"And the message, did she write the message, too?"

"Since any school kid could do it, she had a contest for student input. There was an AI program that stitched it together."

"And the target?"

"That was a little harder to do. There were so many candidates. We basically selected a star not unlike our own with a suitable planetary system. Of course, it needed to be in the right direction for the asteroid deflection. Whether there is anyone there – or, I should say, will be anyone there in a few thousand years – to receive the signal is anyone's guess."

"Well, that's the beauty of art. You do it, not because it's logical, but because it has symbolism."

About the Author

Dev Gualtieri received his PhD in 1974 and had a thirty-five year research career in physics and materials science.

He is listed as an inventor on more than thirty US patents, and on numerous international patents. His eclectic research interests included superconductivity, chemical thermodynamics, magnetism, electronics and computer science. At one time, he was an internationally recognized expert in crystal growth.

Dr. Gualtieri is now retired, and he resides in Northern New Jersey with his wife, Anne. They have a son and daughter who reside with their spouses in Pennsylvania.

Made in the USA
Monee, IL
04 November 2026

67385901R10154